DATE DUE

BRODART, CO.   Cat. No. 23-221

WILD HOPE

WILD
HOPE

*On the Front Lines
of Conservation Success*

ANDREW BALMFORD

*The University of Chicago Press
Chicago and London*

All of the royalties from this book are going to three of the
nongovernmental organizations whose vital conservation work
these stories celebrate.

Andrew Balmford is professor of conservation science in the
Department of Zoology at the University of Cambridge. He is
the coeditor of *Conservation in a Changing World*, and he lives
in Ely, England, with his wife, two sons, and a lot of animals.

The University of Chicago Press, Chicago 60637
The University of Chicago Press, Ltd., London
© 2012 by Andrew Balmford
All rights reserved. Published 2012.
Printed in the United States of America

21 20 19 18 17 16 15 14 13 12   1 2 3 4 5

ISBN-13: 978-0-226-03597-0 (cloth)
ISBN-13: 978-0-226-03600-7 (e-book)
ISBN-10: 0-226-03597-2 (cloth)
ISBN-10: 0-226-03600-6 (e-book)

Library of Congress Cataloging-in-Publication Data

Balmford, Andrew, 1963–
Wild hope : on the front lines of conservation success / Andrew Balmford.
pages ; cm
Includes bibliographical references and index.
ISBN 978-0-226-03597-0 (cloth : alkaline paper)—ISBN 0-226-03597-2 (cloth : alkaline paper)—
ISBN 978-0-226-03600-7 (e-book)—ISBN 0-226-03600-6 (e-book)   1. Nature
conservation.   2. Biodiversity conservation.   I. Title.
QH75.B266 2012
333.95'16—dc23

2011050367

⊚ This paper meets the requirements of ANSI/NISO Z39.48-
1992 (Permanence of Paper).

To Jonah, Ben, and Sarah—for the reason and the rhyme

CONTENTS

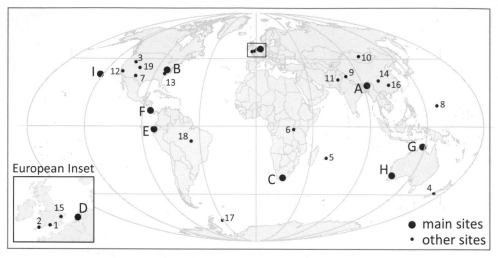

Stories of hope for wild nature: Kaziranga National Park, India (A, chapter 2); North Carolina Sandhills Safe Harbor, United States (B, chapter 3); Working for Water, South Africa (C, chapter 4); National Ecological Network (NEN), the Netherlands (D, chapter 5); Loma Alta, Ecuador (E, chapter 6); PSA Program, Costa Rica (F, chapter 6); West Arnhem Land Fire Abatement project, Australia (G, chapter 6); Alcoa's Huntly mine, Australia (H, chapter 7); American Albacore Fisheries Association, United States (I, chapter 8).

Additional success stories are indicated by numbers: 1, Dorset heathland, UK; 2, large blue butterflies, UK; 3, Yellowstone wolf reintroduction, United States; 4, island cat and rat eradication, New Zealand; 5, Mauritius kestrel recovery, Mauritius; 6, mountain gorillas, Uganda, Rwanda, and Democratic Republic of the Congo; 7, Malpai Borderlands group, United States; 8, Rare radio soap, Micronesia; 9, Spiti Valley snow leopards, India; 10, snow leopards, Mongolia; 11, snow leopards, Pakistan; 12, Central Valley Safe Harbor programs, United States; 13, South Carolina Safe Harbor, United States; 14, Yunnan yew forests, China; 15, Ely Wildspace, UK; 16, giant pandas and crested ibis, China; 17, CCAMLR, Antarctica; 18, Amazon forests, Brazil; 19, Grand Canyon condors, United States.

This map is a Behrmann equal area projection based on information in the *Digital Chart of the World* (Defense Mapping Agency 2001, made available through Penn State University Map Library). Cartography by Ruth Swetnam.

THE GLASS HALF EMPTY

This is intended to be a conservation book with a difference. While most others concentrate on the gloom and doom, my aim is to explore the glimmers of good news. There is no doubt that nature is in grave trouble and that time is fast running out. The year 2010 was the International Year of Biodiversity, during which the world's governments admitted they had failed to meet internationally agreed-upon targets to slow nature's disappearance. But is nature's continued loss inevitable, or are there grounds for hope?

This book tries to answer that question through a global journey in search of places where conservation efforts mean things are getting better, not worse—an attempt to understand conservation success, celebrate it, and learn from it. On each continent I discover what's working and begin to learn why. I find out that while effective conservation sometimes depends on locking nature away in well-protected reserves, other, fresh approaches are yielding positive results too. The key players are no longer just government and conservation organizations; local communities, private landowners, businesses, and consumers can all make a big difference. And although each success story is different, they offer some consistent insights into how projects elsewhere can score more hits and fewer misses, into what ordinary people can do, and about the prospects for wild nature as a whole.

It's a moody-skied day in August, and I'm traveling across the English countryside in the company of an exceptional octogenarian called Norman Moore. Tall and thin with high cheekbones and piercing blue eyes and usually wearing a tweed jacket and tie even in summer, he cuts a striking figure. Norman is one of the founding fathers of the conservation movement and the most knowledgeable naturalist I've ever known. Over a 60-year career, he has helped set up dozens of nature reserves and performed pioneering research on habitat loss and the pervasive environmental side effects of DDT and other pesticides. Along the way he has become a world authority on dragonflies as well as an inspiration to countless young naturalists (my

1

two sons included). Norman is also passionate about heathland—a globally rare type of vegetation more or less confined to sandy soils around the
margins of western Europe, southern Africa and Australia.

Wherever they are, heaths are special. In Britain, they're little patches
of southern warmth—misfits in a muted, edge-of-the-Atlantic climate.
Places where the openness of the vegetation means that (unlike in woods
and wetlands) the sun reaches down to the ground. Places where heathers
paint somberly clad slopes in flamboyant shades of purple and pink. Where
yellow-flowering gorse fills the air with the exotic scent of coconuts, rare
sand lizards and smooth snakes bask on sun-bright banks, and a host of
cold-sensitive birds and insects reach their northernmost limits.

Yet Britain's richest heathlands—the Dorset heaths celebrated as the
hauntingly beautiful backdrop to the novels of Thomas Hardy—have
largely disappeared, victims of over two centuries of plowing, plantation
forestry, and urban sprawl. Where Hardy wrote of bees that "hummed . . .
and tugged at the heath and furze-flowers in such numbers as to weigh
them down" while "in and out of the fern-dells snakes glided in their most
brilliant blue and yellow guise,"[1] there are now housing estates, fast roads,
and featureless fields. By 1960—as Norman documented in the first-ever
study to quantify how quickly people are destroying habitats—Dorset had
lost three-quarters of its heathland.

Stoborough Heath, the focus of much of Norman's early work and where
we're going today, is no exception. With his lanky frame folded in the car
beside me and the green uniformity of the farmscape beneath the scudding
clouds as passing backdrop, Norman recalls Stoborough's singular treasures: its rare Dartford warblers and secretive nightjars, its golden-ringed
dragonflies and emerald wartbiter crickets. Then delight turns to sadness
as he recounts witnessing even this precious remnant being needlessly put
to the plow.

Farming has always been unrewarding on this infertile land, but in the
1960s a government subsidy scheme suddenly meant it made financial
sense to replace the complex and the vibrant with ordered fields of ryegrass
and docile herds of Friesian cattle. The government, concerned about the
country's reliance on imports, compounded the damage by instructing the
Forestry Commission to plant the heaths with pine trees—even though
the soil was so poor it couldn't possibly yield commercially viable timber.

---

1. Hardy, *The Return of the Native*, p. 312. The book is set on Egdon Heath—a fictitious name, but
apparently based on several of the heathland fragments that Hardy knew. "Furze" is an Anglo-
Saxon word for "gorse."

So after years of walking it, marveling at it, and unraveling its intricacies, Norman watched as Stoborough too disappeared. Another downward step on his graph of sustained decline.

This book, however, is about good news. About conservation's successes. About places where a combination of graft, wit, and luck are starting to turn the losses around. But to appreciate how remarkable these stories and the people that have made them are—to fully understand the significance of the hope that they represent—we need first to take stock of where we are and how we got here.

The Emptying Glass

As a conservation scientist, I am deeply moved yet unsurprised by the sad story of Stoborough Heath. There has always been turnover and change in nature—contrary to popular notions, its balance is only ever ephemeral; things move on. And nature is of course resilient too—some species flourish under mankind's influence. But looked at as a whole, the natural world is changing exceptionally quickly, and the overwhelming direction is down. And, just as in Dorset, people are, by and large, responsible.

Since the advent of farming, we've cleared most of the land that's suitable for crop production. Good for us—vital, even—but not so great for the umpteen million other species with whom we share the planet. We've taken over most tropical grasslands, cut down over half of the world's temperate forests, and even converted more than a quarter of the deserts. And habitat destruction is only part of it.

Through overhunting we've reduced the populations of great whales by at least two-thirds, cut wild tiger numbers by over 95 percent, and eaten more than 99 percent of the Caribbean's green turtles. We've compounded the havoc of habitat loss and overkill by moving species to new places where they've variously eaten, infected, or out-competed the indigenous animals and plants. The accidental importation of Asian chestnut blight fungus to New York in the early 1900s unleashed an invasion that killed nearly every adult American chestnut in just a few decades. The rats and cats that European sailors spread around the world's islands are thought to have wiped out at least 35 species of birds. And a suite of introduced mammals—foxes, cats, rabbits, and sheep—are between them responsible for the extinction of at least 18 native Australian mammals.

Species have always gone extinct, of course. The difference is that now our actions have elevated extinction rates to roughly 1,000 times the average, so-called background level seen in the fossil record. At least 1 in 5 of all our fellow species are reckoned to be in danger of extinction in the near future—in some groups, like frogs and corals, the figure is higher still. And our impact is growing.

Since 1970, populations of Africa's spectacular mammals—its elephants, buffalos, lions, and antelope—have halved, and that's inside the parks set up to protect them. Outside, they've often all but disappeared. Over the same period overfishing has seen numbers of most large shark species off the Eastern Seaboard of the United States fall by 90 percent or more. After a decade in which global leaders pledged to significantly reduce the rate at which nature is being lost, the 2010 report card made grim reading. All five measures of the pressures we put on nature were still on the rise, and 7 out of 10 indicators of how much is left showed no letup in how fast we're draining the glass.

Overall, and as a very rough rule of thumb, since the Industrial Revolution people have reduced wild habitats and populations of the species that live in them by around half, and for the past 30 or 40 years we've been removing the remainder at between 0.5 and 1.5 percent each year. The main means by which we're wrecking wild nature—habitat loss and fragmentation, overharvesting, and alien introductions—are well known. But new mechanisms of destruction are emerging too.

People-driven changes to the climate can be held responsible for between a fifth and a third of all species to extinction by 2050[2] and have already caused 1 in every 25 populations of lizards to disappear, unable to cope with us turning up the heat. The catalog of introduced aliens now includes newly described disease-causing organisms like the fungus *Batrachochytrium dendrobatidis*, thought to be responsible for dozens of frog extinctions over the past 30 or so years. Industrial-scale drenching of Europe and North America's forests, wetlands, and farms with nitrates and other so-called reactive forms of nitrogen has caused algae to proliferate; triggered widespread declines of mosses, lichens, and fish; and created immense oxygen-starved "dead zones" in coastal waters.[3] And extraordinarily,

2. Conservation scientists use phrases like "commit to extinction," because even though they cause species to disappear, the impacts of habitat degradation and climate change are rarely instantaneous. In ecologist Dan Janzen's phrase, some individuals hang on as "the living dead," with populations persisting—albeit in ever-diminishing numbers—for maybe a few decades or even a century. Others more euphemistically refer to this lingering death as "relaxation."

3. Although as a gas nitrogen makes up 78 percent of the earth's atmosphere, its reactive forms (such as nitrate and ammonium ions—on which all life depends) are much scarcer. The early-twentieth-century development of the Haber-Bosch industrial process for manufacturing ammonia (and from that, synthetic fertilizers and explosives) enabled Germany to fight World

our emissions of carbon dioxide—a quarter of which are absorbed by the sea, where they form carbonic acid—are now on such a scale that they're shifting the pH of the oceans. By 2050 the water may be too acidic for many creatures to build their calcium carbonate shells. Much marine life—from reef-building corals to photosynthetic plankton—might quite literally dissolve. We face the prospect, as marine biologist Jeremy Jackson puts it, of a world without seashells. Try explaining that to the grandchildren.

So why is it all happening? Why is one species—our own—in the process of precipitating an extinction spasm of a magnitude not seen since the last mass extinction event 65 million years ago, when an asteroid struck the Yucatán Peninsula in Mexico, wiped out all the dinosaurs that hadn't evolved into birds, and ushered in the age of the mammals? The underlying causes of today's crisis—the drivers, in the jargon—fall in four main groups. Most obviously there's the size of our population—which took almost all of human history to reach the billion-mark (sometime in the early 1800s) but which, staggeringly, has recently been growing at around 1 billion people every 12 years. That's equivalent to a new Athens- or Nairobi-sized city every month. Growth is now slowing, with world population probably peaking at 9 to 10 billion in the second half of this century, but many argue that's still many more than one planet can sustainably support.

Second, there's our unquenchable demand for higher standards of living—essential for much of the world's population, yet far more questionable among the rest of us. The numbers show that this is probably an even bigger factor—though less comfortable for the comfortably off to contemplate—than population growth. Humanity's combined demands on the planet can be thought of as population size multiplied by per capita consumption. Yet while total population is likely to rise by roughly 50 percent between 2000 and 2050, per capita incomes are forecast to grow more than threefold—so their effects on individual consumption are likely to far outstrip those of growth in the total number of people doing the consuming.

Next on the list is intrinsic human selfishness. When we make choices we tend to put ourselves above people elsewhere and above future generations. This is bad news for conservation, because the benefits of conserv-

---

War I and has doubled the number of people that farming can feed. But the resulting surge in nutrient levels (termed "eutrophication") from fertilizer and sewage runoff as well as fossil fuel combustion has greatly harmed species adapted to low nutrient conditions and created vast oxygen-starved dead zones in lakes and coastal waters (one in particular in the northern Gulf of Mexico is now the size of New Jersey). According to a recent authoritative review the reactive nitrogen bonanza also has increased ground-level ozone, airborne particles and associated human respiratory and cardiovascular diseases, and reduced the life expectancy of half of all Europeans by an average of six months.

ing somewhere—say, of keeping a wetland as it is rather than draining it for agriculture—are often what economists term externalities: they accrue mostly to people other than those in charge of it. Downstream villagers may gain from the clean water the wetland provides, distant naturalists may feel happy that its rare birds continue to thrive, and so on. But because these benefits are not experienced by the would-be farmer deciding whether to drain the wetland, they'll tend to be ignored. And by the same token, because the benefits of conservation often build up only over the long term, people will typically discount them—not just in their heads but on their balance sheets—in favor of more immediate returns. Our narrow, short-term decision making generally penalizes the rest of the planet.

The last root cause is our growing disconnect from nature. We live in a rapidly urbanizing society, where for the first time more than half of humanity now works, plays, and sleeps in towns and cities: no longer immersed in the natural world and attuned, for our own survival as farmers or fishers, to its patterns and rhythms. Instead we spend our lives indoors, in cars, and online in places like Brooklyn, Bangalore, and Brussels. As a consequence, many argue, we're losing touch with wild creatures and wild places. We can no longer tell our lady's mantle from our lady's slippers, our frogbit from our froghopper. We no longer know what phase the moon is in, let alone how high the next tide will be. And there's the problem. How can we be expected to care about what we no longer experience, what we no longer know? Nature's erosion may ultimately be driven as much by our indifference as by our direct actions.

Yet wherever we live—however removed—the current collapse of the living world affects us all. For many people there is a fundamental moral argument that says such loss is simply unacceptable. Some express it in religious or spiritual terms. Some are motivated by the realization that all living things are related to us, that we are family. The distinguished photographer and biologist Roman Vishniac once said, "Every living thing is my brother. How wonderful that is." Others are driven by a sense of duty to hand the world on to future generations in no worse a state than we found it. Theodore Roosevelt summarized this argument when he wrote, "The nation behaves well if it treats the natural resources as assets which it must turn over to the next generation increased and not impaired in value."[4]

For me, alongside respect for relatives and responsibility to be good custodians, there's another motivation: a sense of wonder in nature's marvels—whether that's in witnessing a cuttlefish change colors almost

4. From Roosevelt, *The New Nationalism*.

instantaneously as it glides over the kaleidoscopic busyness of a rockpool; or in learning about the extraordinary life of eastern Queensland's gastric-brooding frogs, which swallow their eggs (to protect them from predators) and then develop them inside their stomachs;[5] or in getting my gravity-bound brain around the notion that the common swift fledglings that take their first flight in my garden each summer will literally not touch ground again (not to eat, rest, or even to sleep) until they themselves return to nest as two- or three-year olds. Nature is jammed full of such wonders, and what makes me an ardent conservationist is the desire that my children and the generations that come after them can have their own opportunities to be enticed, amazed, and humbled.

I appreciate I may be a little unusual, but for those who might be less moved by the moral or aesthetic case for conservation there are powerful material arguments. We all gain from what are now labeled ecosystem services—benefits provided for us, for free, by nature. The problem is that like most things that we get for nothing, we often overlook these services until there's a crisis. A canopy of trees can protect hillside soils from erosion, and wetlands can store immense volumes of water. Large predatory fishes often keep in check smaller predators that might otherwise eat things we want for ourselves, and wild scavengers dispose of dead animals safely and quickly. But they all do so unseen and unnoticed.

Until, that is, mudslides in the Philippines remind us of the perils of deforestation, and the dreadful impacts of Hurricane Katrina illustrate what happens when we drain swamps that normally buffer people from storms and floods. Until the removal of those east coast sharks triggers a population explosion among the rays they used to feed on, which in turn chomp so many scallops they destroy a century-old fishery in Chesapeake Bay. Or until a newly available veterinary drug inadvertently poisons tens of millions of southern Asia's vultures, leading to a buildup of cattle carcasses and grave fears about outbreaks of disease and escalating numbers of feral dogs.[6]

5. . . . or did, until both species disappeared in the mid-1980s, most probably because we introduced that frog-killing fungus to the forest streams where they lived. Tragically, the details of how they turned off the production of stomach acid and so nurtured rather than digested their young—and the insights this may hold for medicine—will therefore remain forever only partly understood.

6. Vultures were until recently so common across the Indian subcontinent that in places they were considered a hazard to aircraft. Then in the 1990s populations plummeted—by as much as 99 percent in just ten years. The mystery of their sudden disappearance was finally solved when it was realized that diclofenac—an anti-inflammatory drug that causes kidney failure in some birds—had recently come out of patent and was being widely used to treat sick cattle, whose

But crises aside, even in our everyday lives we all rely on wild places and the creatures that live in them. Despite falling stocks, half of the fish we eat is still caught in the wild. A third of all crop production depends on pollination by animals, many of them wild insects. And the great planktonic soup of the world's oceans (much of it living in those pH-sensitive shells) helps stabilize our climate, not just by soaking up carbon dioxide through photosynthesis, but in some cases by directly reflecting sunlight back into space and even releasing into the air particles of a chemical called dimethylsulphide, which in turn breaks down into tiny sulphate droplets around which clouds then begin to form.

Given the myriad threads of our dependence on wild nature, it's no surprise that the damage we're inflicting on it is having a significant impact on people. The supply of wild-caught food is going down, so too are clean water and wild-harvested timber; pollinators are declining, as are populations of many birds, frogs, and insects that help control pest outbreaks on farmland. As shrinking natural habitats become less able to shield us from storms and cushion us from climate change, insurance companies are passing on the escalating costs of flood damage by raising insurance premiums.

A recent, UN-backed initiative called TEEB (The Economics of Ecosystems and Biodiversity) argues that the impacts of nature's erosion are especially severe for the world's poorest people, who rely on ecosystem services for up to 80 percent of their incomes. The report's authors warn that unless such benefits are valued and factored into economic decision making, continued environmental degradation will risk livelihoods, lives, and even the global economy. Many fear that if we carry on this way, living on the ecological never-never, spending down the planet's natural capital with little regard for when or even if it can be repaid, then the recent turmoil caused by overspending our financial capital may come to seem like a minor hiccup.

In material as well as moral terms, then, our ongoing erosion of nature really matters. Yet faced with the global litany of loss and our manifest ability to ignore the stark warnings of its consequences, one could be forgiven for submitting to a sense of hopelessness. Of giving up. Of deciding that, essential though changing the way we treat the rest of the planet may be, our

---

bodies were in due course eaten by vultures. Diclofenac manufacture is now banned in three countries and an alternative, nonpoisonous drug called meloxicam is being promoted in its place, but recovery of the vultures and their free waste disposal service will take many decades.

behavior is simply too entrenched and the momentum behind the drivers of nature's demise simply too great.

I confess to feeling exactly that way when a developer bought up one of the last few fragments of nature in Ely, the tiny city where I live. Since being abandoned by industry, a flooded clay pit had become home to shy otters and nationally threatened bitterns.[7] The new owner promised to convert it into a residential marina. Local politicians backed his proposal as being good for jobs. Planners seemed reluctant to risk legal action and talked of development and compromise. People who treasured the area as it was began to despair, to believe there was nothing they could do. And perhaps that makes sense. Maybe nature's loss is inevitable and the best course of action is to hide under the covers. Globally as well as locally, maybe conservation is simply too difficult.

A Fragment Reprieved

Yet back in Stoborough the story has a rather more positive twist. As Norman and I draw close to the once-wild place he witnessed being destroyed, in place of the expected regularity of fields and ordered ranks of pine plantations I see the olive and purple of heathers and the perennial yellow of the furze. We park, Norman unpacks his 80-something-year-old legs, and we start to wander a recreated patchwork of habitats: not dull, managed uniformity, but wildness once more. As if on cue, the sun breaks out from behind a cloud and the colors sharpen. Within minutes we're watching a rare Dartford warbler, tail cocked and crest raised, singing defiantly from a gorse bush. A female sand wasp carrying a paralyzed caterpillar quarters the bare ground ahead of us, searching for the hole where she will lay her egg and stock its larder with her prey. Nearby, white tufts of cotton grass brighten the fringe of a small bog, where sticky-leaved sundews supplement the meager rations they draw from the soil by trapping small flies. Norman is soon reeling off the scientific names of passing dragonflies like

---

7. Great bitterns are near-mystical, reed-dwelling herons. The deep booming calls given by breeding males are the source of more than 20 colloquial names in English—including bog-blutter, mire drumble, buttle bump and bull of the bog. Although the calls can carry for miles, the birds' shyness, magnificent camouflage, and sensitivity to wetland drainage make them famously difficult to see. By 1997 there were only a dozen booming males left in the whole of the UK, hence many people's outrage at the prospect of converting bittern habitat into parking spaces for barges.

other people talk of old friends. He sees my enjoyment too, and smiling, explains the reason for Stoborough's dramatic turnaround.

"Back in the 1960s, while the government was urging its foresters to plant uneconomic pines, I made friends with a sympathetic forest officer. He sent me information about which tracts were due to be planted, and I was able to persuade him to spare some of the most precious bits of the heath." A holding operation: conservation by stealth. Fast-forward another 20 years, though, and attitudes had changed completely.

"The government had by now decided that public access and enjoyment were more important priorities than unprofitable plantations, and so its foresters began working with conservation organizations to turn large areas of conifers back into heathland." Trees were removed and the sunlight let back in. Seeds, long dormant in the seedbank, started geminating, and clever ecologists learned how to help them on their way. Birds, insects, and reptiles spread out from the nuclei that Norman's foresight had saved. The dizzying, intricate machine of heathland began working once again.

Encouraged by an unprecedented alliance of conservation organizations, local and European governments, foresters, firefighters, and even the UK Ministry of Defence (which owns much of the land), a similar process of renewal has since got underway across a large swath of what was Hardy's great Egdon Heath. Altogether, some 1,200 hectares—around one-sixth of what is left—have been restored by clearing scrub and cutting pines. Reinstating traditional grazing keeps the renewed heathland open and mimics what wild grazers and wildfires would once have done. My friend Norman, the first man to chart the decline of a habitat and one of the first people brave enough to speak out passionately against it, has lived long enough and worked hard enough to see the curve of loss just starting to tilt upward.

And despite the global gloom, Stoborough is not the only bright spot, of course. Not far down the road in Somerset, pioneering ecologist–cum–insect detective Jeremy Thomas has painstakingly rescued the large blue butterfly from nationwide extinction. This dazzling steel-blue species disappeared from Britain in 1979, a casualty of the cessation of traditional sheep grazing and the resulting encroachment of scrub into its grassland habitat. The key to its successful reestablishment (using insects from Sweden) lay in teasing apart the details of its extraordinary life cycle. It turns out that the butterfly not only has to lay its eggs on wild thyme (itself a rather picky plant), it also needs a nearby colony of a very particular species of red ant (a beast by the name of *Myrmica sabuleti*). Why? Because this innocent-looking animal is in fact an obligate ant-killer.

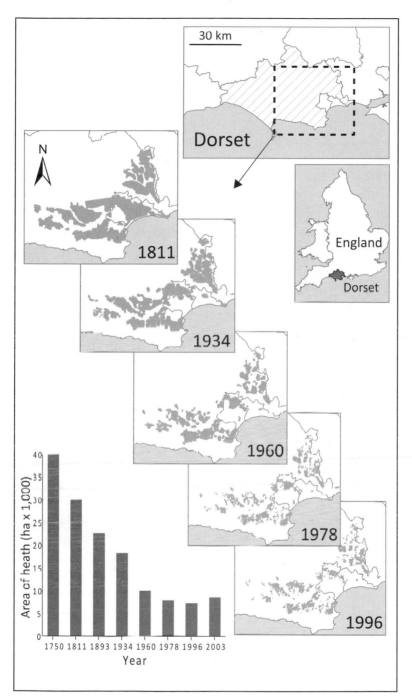

Map 1. The fall and eventual rise of Dorset's heathlands. The maps from 1811, 1934, and 1960 are from Norman Moore's seminal 1962 study. More recent data are from Rose et al. 2000 and Cynthia Davis, Kathy Hodder, Rob Rose and Nigel Symes (personal communication). The graph shows the total area of heath in thousands of hectares. Cartography by Ruth Swetnam.

After a few weeks of feeding on thyme the caterpillar of the large blue drops to the ground and gives off a secretion that attracts its preferred ants, which mistake it for one of their own larvae and transport it underground into their nest. In a remarkable display of filial ingratitude the caterpillar then proceeds to gorge itself on ant grubs, all the while still being protected by its unsuspecting hosts. After fattening up, hibernating and pupating in the nest the butterfly emerges as an adult, and the cycle begins again.

Remove the particular species of ant that it exploits, though, and the large blue cannot cope. And to get enough warmth in Britain's soggy northern climate, the ant requires grass trimmed to 3 to 5 centimeters in height, so without just the right amount of sheep grazing, both the ant and its beautiful parasite disappear. Complicated. But when Thomas and colleagues finally unraveled all the facts and got the grazing spot-on in places that still had the right ants (not forgetting the caterpillars' thyme), the reintroduced butterflies flourished. And once the team had cracked it, they scaled up, so that now on more than 30 hillsides across the southwest the glorious, treacherous blue ant-killers are back.

There are other successes too. The 1995 reintroduction of wolves to Yellowstone National Park in the northern United States raised hackles among local landowners, but is starting to deliver a broad sweep of ecological benefits. Populations of coyotes (which are common and don't get on with wolves) have decreased, and survival of young pronghorn antelope (which are rare and don't get on with coyotes) has increased. By providing a steady supply of carcasses through the winter, wolves may inadvertently be helping scavengers like bears and ravens cope with climate change reducing the number of animals dying from cold. And with wolf numbers going from zero to over 100 in 15 years, elk populations have declined and moved into dense conifer forests where they're less vulnerable to attack. Some ecologists have even suggested that the resulting relief from years of heavy elk browsing may be driving a recovery of stands of willow and aspen (and the songbirds, small mammals, and beavers that depend on them).

Around the world's oceanic islands, serious beefing up of efforts to eradicate introduced predators—involving a new generation of poisons and militarily efficient bait delivery—has seen the successful removal of alien rats from land masses up to 100 square kilometers in area. Feral cats have been completely cleared from islands 3 times that size. Many bird species threatened by the alien invaders have benefited, but so too have a host of reptiles and insects found nowhere else, such as New Zealand's extraordinary tuataras (the sole survivors of an ancient reptile lineage otherwise known only from fossils) and wetas (flightless crickets that can weigh as much as mice).

On Mauritius, once home to the dodo, a heroic program to save the island's kestrel—brought to the brink by DDT and introduced predators and believed by many to be following its fellow Mauritian into oblivion—has restored the population from just 4 individuals in 1974 to over 800 adults today. And globally, a careful analysis of population trends, threats, and management actions has identified 30 other still-surviving bird species that would almost certainly have gone extinct without conservation interventions, such as restoring habitat, controlling predators, or boosting bird numbers through breeding in captivity. Most of the species are still dependent on conservation action, many of them are still in intensive care, but all, nonetheless, are with us still.

So in spite of the daunting enormity of the challenges, conservation projects can and do work. And in looking on the bright side for a change, I was reminded of my half-is-still-left rule of thumb: over the 250 years since the Industrial Revolution, on average we've roughly halved the populations of wild species and the areas of habitat where they live. The glass half empty, for sure, but despite everything we've done, half full, too. Contrary to what we might think, more than half the tropical rainforests still stand. Most coral reefs remain. More than a billion people notwithstanding, India continues to be inhabited by leopards and bears, crocodiles and wolves. Despite grinding poverty, rapid population growth, and the dreadful chaos of civil wars, the Great Lakes region of Africa still has lions, elephants, and more mountain gorillas than when I was the age my sons are now. A great deal, of course, has been lost, but there's also a great deal left to fight for.

I began asking myself whether, within the conservation movement, our own pessimism might be part of the problem—not just a symptom but an unwitting accomplice to decline. In trying to ensure that policy makers and the public at large appreciate the seriousness of the problem, maybe we've focused too much on the negative. In one way it's worked—I think most people nowadays have got the message. They know that the natural world is in grave trouble, and that something must be done. But what? Perhaps in selling the bad news quite so effectively, we've overlooked the vital importance of believing there are solutions. We've made people painfully aware but given them no prospect that things can be turned around—just a dismal choice between despair and denial.

Good News for a Change?

The issue started to nag at me. I began to look at other sectors struggling to make a difference, and realized maybe we were missing a trick. Take Red Nose Day—a major UK telethon that raises around US$100 million every two years for humanitarian projects both at home and across Africa. Like many people I don't think much about it until on my way home that evening an otherwise presumably sane adult dressed up as a rabbit and holding a bucket accosts me for my small change. But a couple of hours of variously amusing TV sketches and a heartrending plea or two later and I've gladly picked up the phone and pledged the family silver.

Poverty is of course every bit as vast and pressing an issue as the collapse of biodiversity, with the same potential to appear utterly overwhelming, so how does Red Nose Day make millions of us open our wallets so readily? It helps no doubt that the show is fun, but more than that, I think it works because it immediately follows up desperately moving films about AIDS victims in Kigali or homeless teenagers in London with uplifting sequences showing where my money went last time. Showing me something can be done, that there are successes, that I can make a difference, right now, without even leaving the comfort of my sofa. Sorrow, guilt, and redemption, all in the space of five minutes.

Maybe conservation could gain from dwelling a little more on its successes—from celebrating them, learning from them, and using those insights to encourage and enable people to do more. I wondered whether in my own city a more positive outlook might help save the wetland oasis I cared about from being turned into a watery caravan park. And more broadly, maybe a greater understanding of instances where conservation is working could increase the chances of future triumphs and help build momentum for greater action.

I dug further, asking friends and colleagues around the world to send me their stories of success. They often pointed out what we all knew—that bad news stories are far more common, that setbacks are the norm. But they sent me nuggets of good news too. At first a handful, and then many more, often featuring novel interventions and unexpected players. I learned that in the borderlands of Arizona and New Mexico independent-minded cattle ranchers are cooperating to improve their grazing by reintroducing long-suppressed wildfires. But they're going further too—actively encouraging once-persecuted species, such as prairie dogs, endangered rattlesnakes, and even jaguars, back onto their land. Across the far-flung islands of the western Pacific, where an innovative conservation NGO called Rare broadcasts a popular radio soap opera dramatizing social and environmental issues, a quarter of listeners report that it's prompted them to stop eating sea turtle eggs, and almost a fifth say they've adopted family planning as a result of the show.

In the remote Spiti Valley in India's Trans-Himalayas herders who were losing more than one-tenth of their livestock each year to threatened snow leopards have enrolled in an insurance program. As long as they no longer persecute large carnivores, this offsets their losses and includes additional financial rewards for community herders taking closest care of their flocks. Three villages have also created livestock-free reserves where wild sheep numbers are recovering—a boon to local snow leopards. Meanwhile, in Mongolia, herders helped to produce and sell knitwear for foreign markets have in return stopped hunting snow leopards and their wild prey and have seen their incomes rise by a quarter as a result. And in Pakistan a much-needed vaccination program addresses the diseases that are the biggest cause of livestock mortality in exchange for villagers ceasing snow leopard persecution and reducing the size of their flocks (so that there's more grazing for wild herbivores, which snow leopards can then hunt).

I eventually cataloged scores of good news stories, but the more I learned, the more curious I became. Are these examples truly successful, and if so, why? What's made them work, and what does this reveal about how our ideas about conservation are changing? And most importantly, what does understanding some of its successes tell us about the prospects for conservation into the future? Is nature's demise inevitable, or could enough be done, soon enough, to make a difference?

I realized that to get answers to my questions I needed to see the projects firsthand, to meet the people who've made them happen, find out their stories and frustrations, and judge their success for myself. But not wanting to spend too long away from my family, growing awareness of my carbon footprint, and the small matter of a full-time job meant I couldn't visit them all. I needed to prioritize.

To try to make what would have to be a small and hence unavoidably unrepresentative sample as diverse as possible, I deliberately picked projects that differed in the sorts of species they targeted. Because I wasn't interested in whether conservation could work in just one place, I made sure the list covered a varied set of countries. I looked for local interventions as well as large programs, species and places facing diverse threats, and projects that tackled them using very different sorts of approaches.

Because I was interested in wrinkles and complications, none of the projects I chose have been unequivocal successes: all have faced setbacks and some may yet be reversed. And to avoid upsetting some of the conservation organizations I usually work with by omitting their flagship successes, I did my best to avoid all of them, picking only those projects I knew little about before I began. Eventually I whittled the long list down to just seven—one from each continent and one from the sea[8]—and set out to visit them one by one. This book is my record of those journeys.

In each place I came face to face with extraordinary animals and plants and met the heroes and foot soldiers who are trying to save them. Though each good news story was complex and still a work in progress, I learned that conservation successes are widespread and diverse. I found that, despite the rhetoric of this or that NGO[9] or academic, one size doesn't fit all, and different problems call for different solutions. Most of all, I learned how real progress often hinges on being able to see problems not just through a conservation lens but from the perspective of society as a whole. How do local farmers or the government or a major corporation perceive the issue, and can their concerns be turned into an opportunity for conservation rather than a threat?

My journey starts in the floodplains of Assam, where dedicated rangers and exceptionally tolerant villagers have between them helped bring Indian rhinos back from the brink of extinction. In the pine forests of the Carolinas I discover why plantation owners came to resent rare woodpeckers—and what persuaded them to change their minds. In South Africa I learn how invading alien plants have been drinking the country dry and how the Southern Hemisphere's biggest conservation program is now simultaneously restoring the rivers, saving species, and creating tens of thousands of jobs.

The conservation problems I meet are diverse, and the solutions people have developed become progressively more innovative through the book,

8. I confess this list excludes Antarctica—not because I forgot it, because it doesn't face threats (it certainly does), or because it lacks successful conservation projects (I mention one briefly in chapter 9), but simply and rather feebly because by the time I'd sailed there, researched the story and sailed back, even the superhuman patience of my editor would have been stretched beyond its limit. That, and I'd have got very seasick.

9. Conservation is unfortunately awash with acronyms. I've done as much as I can to keep them at bay, but they keep coming back. NGOs are nongovernmental organizations (such as Environmental Defense Fund [EDF] in the United States, and the UK's Royal Society for the Protection of Birds [RSPB]). While I'm on the subject, IGOs are intergovernmental organizations (like the International Union for the Conservation of Nature and Natural Resources [IUCN]); and Big International NGOs (like The Nature Conservancy [TNC], Conservation International [CI] and the World Wide Fund for Nature [WWF]) are sometimes called BINGOs.

from relatively conventional approaches involving reserves and guards (the so-called fortress-and-fines model) to more recent and radical ideas like limiting the unintended downsides conservation can impose on landowners and using carbon taxes to reward forest conservation rather than clearance. In the Netherlands I find local government officials paying for ecological restoration on an almost imaginable scale. In Ecuador the answer involves peasant farmers opting to save their forest because of the impact its loss is having on the flow of water to their fields. In Australia I find out why the world's largest aluminum-mining operation is spending a fortune growing eucalypts is and dropping millions of poisoned salamis from high-tech planes. And in my final story I discover how ordinary consumers on the other side of the world can make a real difference to mid-Pacific tuna and the West Coast fishermen whose livelihoods depend on them.

These stories, seven in all, follow a thread from traditional conservation through to mold-breaking innovations and build to the final chapter, where I examine what the projects have in common, and an appendix, where I suggest ten things that readers themselves can do to make a difference. Each story also offers specific lessons on how to win more conservation battles. Together they yield common themes, too—elements that crop up time and again. But for me, most of all they offer one particular, overarching insight: the key ingredient the conservation movement is in danger of forgetting, and a means, perhaps, to begin to win the war. Hope.

In conservation, we immerse ourselves in the bad news. We document it, we cleverly investigate its causes and consequences, and we find ever more compelling ways to communicate it to the outside world. We need to. I've no doubt that conveying the stark realities facing nature is vital. But most people hearing that message still don't translate it into changes in their own lives. I've come to think that's in part because they believe that the continued erosion of the living world is an unavoidable downside of the human enterprise. Faced with an unending barrage of bad news, they see no chance for turning things around.

This book is my attempt to challenge that hopelessness.

GUARDING THE UNICORN:
CONSERVATION AT THE SHARP END

The century-long armed struggle to save Indian rhinos in the Northeast Indian state of Assam illustrates how traditional but unfashionable fortress-and-fines conservation can sometimes work and in places may even be essential. But it also raises awkward questions about the negative impacts of conservation, and about whether—as the rest of the book explores—conservationists can devise alternative approaches that impose less of a burden on local people.

My quest to seek out conservation's successes begins 3 meters above the ground, swaying across a mist-bound sea of grass on the back of a freckle-eared, 40-year-old mother of three called Mohan Mala. She's one of a herd of working elephants that spend each dawn transporting groups of wide-eyed tourists into Kaziranga National Park in Assam.

A few minutes into our journey the sun edges into view, coloring the neutral grays of the landscape with greens and yellows, and picking out dew-beaded threads of spiders' webs. Dense elephant grass extends all around, 4 or 5 meters high. A small group of wild boar, bristly black and curious, stops to sniff at us then trots busily off. But even from our lofty vantage point, it's hard to spot animals in the towering green sward. It's quiet, too: the mist muffles the few birds that are starting to sing, and the elephants walk almost noiselessly.

People come here from around the world for close-up encounters with one of the rarest animals on the planet. Most of the passengers on this morning's elephant convoy are Indian families—children squeezed between mums and dads, grandparents and uncles, all chattering eagerly about what they hope to see. But today I reckon I'm the elephants' most excited customer. Despite the early start, I woke well before the alarm, nervous with expectation: tracking down Kaziranga's most famous inhabitant is something I've longed to do since I was a child.

The elephants descend with surprising grace to cross the muddy edge of a *beel*, a shallow pool. I glance down at the enormous footprints our convoy is leaving behind—small craters, bigger than dinner plates and inches deep. The line of giants slows as it climbs the other bank, then snakes for a while longer through the shrouded wetness of the grassland. We reach a second *beel*, and suddenly, there it is. Belly-deep in the mud, peering shortsightedly at us. A great armor-plated beast, almost as vast and more extraordinary still than the elephants: an Indian rhino.

Battleship gray, with huge flank folds and backside creases and studded with wart-like bumps on its legs and shoulders, *Rhinoceros unicornis* (named for its single horn) is the species most people have in mind when they think of a rhinoceros. I'm taken back to my first images of this animal—a four-inch plastic replica I adored as a child and photos in a wildlife book I used to treasure that still sits, dog-eared and broken-spined, on my shelf. Later, the poster on my wall of Albrecht Dürer's famous woodcut that so captivated sixteenth-century Europe—even more exaggerated in its armored splendor than the real thing. And memories of my young children's astonishment at seeing Indian rhinos in the zoo—one animal so gigantic, even by rhino standards, that the keepers had had to enlarge the door to its house.

Still, nothing has quite prepared me for coming to within a few muddy steps of this immense beast in the wild. This one's a male. He's nearly 2 meters at the shoulder, around 4 meters long, and weighs about the same as a couple of family cars. For a time he looks up at us through the morning dampness. We stare back, our wonder tempered with a slight feeling of apprehension: I've seen what's left of a national park truck that one of his friends took exception to. But apparently content that we're mere protrusions on some rather misshapen elephants and not a threat, he snorts, settles back to his wallow, and ignores us.

With their thrilled clients beaming broadly, the mahouts perched up on the elephants' heads to steer them signal to their charges, and we move on. We make our way steadily across a broad plain. As the mist thins, steel-blue hills appear in the distance. A patch of shorter grass, waist-high, opens to our left then bursts into movement as a herd of swamp deer startles. They bound up, russet-brown leaping through the green, and then regroup before warily crossing our path and returning to their resting place.

We carry on. Our guides have spotted a pair of brown-tufted ears, golden-edged in the morning light. Another rhino, this time a female, with a young calf in tow. As we approach, they trundle into a pool, the mother keeping a watchful eye on the early-morning intruders. Cameras click while Mohan Mala and the rest of her team stand patiently by. It's the same at the next waterhole: another female, and another youngster—this one only a few months old, with the tiniest nub of a horn.

Kaziranga is one of the best places in the world to see rhinos. After the elephant ride is over, I take a long journey into the park and count no fewer than 58 scattered along the shore of one particularly large *beel*. All told, there are around 2,000 lumbering about the place. But the truth is they came within a hair's breadth of disappearing completely. When conservation efforts first began in the area at the start of the twentieth century you could have tallied all the remaining rhinos just on your fingers and toes. Less than a score remained. Since then, a hundred years of painstaking efforts have increased the population a hundredfold: two-thirds of all the world's Indian rhinos now live in Kaziranga. Yet to me, the idea that there are any of these spectacularly rare animals here at all, in one of the poorest, most densely settled, socially complex areas and politically fraught places on earth, is little short of miraculous.

Unlikely Survivors

Rhinos the world over have long had a hard time at the hands of people. Roughly half of the rhino species that were around when modern humans first got going about 130,000 years ago—and which roamed not just Africa and Asia but Europe too—have since disappeared. The same is true of most of the rest of the so-called Pleistocene megafauna—the huge-bodied and now extinct mammoths, cave bears, and giant deer of Eurasia; the elephant-like mastodons, enormous ground sloths, and armor-plated glyptodonts of the Americas; Australia's giant marsupials; New Zealand's flightless moas; and the even larger elephant birds of Madagascar. Because these extinctions happened for the most part many thousands of years ago, the exact causes are still disputed.

Some scientists argue that the climatic upheavals of the last ice age were responsible. But others point out that these spectacular creatures had survived many previous cycles of glaciation and warming. What's more, they disappeared at very different times in different places—out of step with the changing climate but suspiciously soon after people first colonized their respective landmasses. The speed of extinctions varied too. They happened quickest in places (like North America) that were suddenly exposed to sophisticated human hunters. If the megafauna survived at all it tended to be in much longer-colonized places (like Africa and Eurasia) where early hunting was presumably less advanced and developed only gradually, perhaps allowing for the evolution of adaptations against predation by people.

Other evidence—like the discovery of the butchered remains of some of these animals in human middens—has led many researchers to conclude that people and not climate were probably responsible for the great majority of extinctions of the megafauna. The relative importance of overhunting and of humans destroying habitats is still unclear, but most scientists now agree that one way or another people played a large part.

Five species of rhino did survive this prehistoric purge—the white and black[1] in Africa and the Indian,[2] Sumatran, and Javan in Asia. All are now in varying degrees of trouble—victims first of habitat loss and, more recently, spiraling demand for their horns. In Africa, the southern population of white rhinos, which plummeted to less than two dozen in the early twentieth century, now stands at a healthy 17,000, but the northern race has in all likelihood recently become extinct in the wild, a casualty of the long-running unrest in the Democratic Republic of Congo. The black rhino, once the most numerous species, collapsed from around 70,000 in the 1960s to 2,500 just 30 years later and stands at around 4,000 today. The Javan rhino hangs on in a single isolated population of less than 50 animals. And the small, forest-dwelling Sumatran rhino numbers less than 300, scattered across the remaining forest fragments of Borneo, Sumatra, and Peninsular Malaysia.

Historically, the Indian rhino ranged across a great swath of the subcontinent, from Pakistan through India, Nepal, Bangladesh, and Bhutan, right to the borders of China and Burma. All three Asian species occurred in Assam. But now the Sumatran and Javan rhinos are confined to Southeast Asia, and the Indian species survives in just 10 or so protected areas—three in Nepal and the rest in northern India. The main historical reasons behind the decline were centuries of habitat loss to make way for farming, combined with sport hunting by powerful elites. Toward the end of the nineteenth century one particularly keen British colonel—a Fitzwilliam Thomas Pollock—singlehandedly killed 47 rhinos, while the Maharaja of

---

1. Despite these names, all rhinos except the reddish-haired Sumatran species are gray. The label "white" was derived, mistakenly, from the Dutch word *wijd*, describing the broad, lawnmower-action lips of this grazing specialist; "black" was then used to distinguish the second African species, whose mouthparts are instead adapted for browsing. "Square-lipped" and "hook-lipped" are therefore more accurate names for these species, but take a lot longer to type.

2. Given its distribution and the testy relations between the countries it inhabits, the Indian rhino is sometimes more diplomatically called the greater one-horned rhino.

Cooch Behar in what is now West Bengal reportedly succeeded in dispatching no fewer than 207.

By the turn of the twentieth century it was clear that time was running
out, and Assam was the only place in India where the species could be saved.
Anxious letters were exchanged between senior Raj administrators. After
apparently finding only footprints on her trip to spot one of the last few
rhinos, Lady Curzon, American wife of the viceroy, added her influential
voice to the calls for protection. She asked her husband to intervene, and
by 1905 the area around Kaziranga was provisionally notified as a reserved
forest—one of the world's first formally protected areas.

Since then, the protection formally afforded Kaziranga has increased

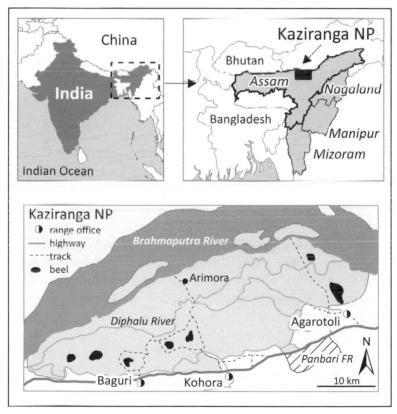

Map 2. Kaziranga National Park lies in the floodplain of the Brahmaputra—"the naughtiest
river in India" and a dominant force in the lives of the creatures and people that live there.
Map based on information given in Barua and Sharma 1999. Cartography by Ruth Swetnam.

a good deal—with its designation being upgraded to a game sanctuary, a wildlife sanctuary, and in 1974 to a full-blown national park. The threats the area faces have changed too. With the countrywide banning of rhinoceros killing in 1910, hunting for sport dwindled, but since the 1960s illegal killing of rhinos for their horns has increased dramatically. Unlike the African species—whose horns are also in demand for making handles for traditional Yemeni daggers or *jambiya*—the horns of Asia's rhinos are used almost exclusively in powder form in traditional Chinese medicine.

Contrary to reports in the Western media (and even though the adjective for rhinoceros-like is rhinocerotic), the main medical use for rhino horn is not as an aphrodisiac. Instead, it's used primarily as a fever-reducing agent, and there is some clinical evidence that it does have slight effects, though no more than water buffalo horn. With the growth in population and particularly wealth of China and its neighbors, the price of rhino horn has soared. By the time it reaches the market in China and Hong Kong, a kilo of Asian horn can fetch over US$60,000—substantially more that the same mass of gold. The resulting incentives to poach the few rhinos left alive are obvious.

In contrast to the threat from poaching, human-caused habitat loss is less of a problem. Kaziranga protects just over 400 square kilometers of grassland, forest, and swamp, and there's been limited encroachment into the park by settlers. One of the neighbors does destroy habitat, however: the mighty, rapidly shifting Brahmaputra. One of the largest rivers in the world not yet under the control of major dams, the "son of Brahma" bounds the northern edge of Kaziranga and dominates life there. Between June and September, the great monsoon-swollen river bursts its bank and spills out southward across the wide, flat grasslands that cover most of the park. The floodwaters sweep right up to Kaziranga's southern boundary, where tidily kept tea estates blanket the base of the Karbi Anglong Hills. Meanwhile, the surging river washes away great chunks of ground, literally eating up the park. Since the 1960s, Kaziranga has lost around 50 square kilometers of dry land to the forces of erosion. The river does give back what it takes—as the main channel shifts, new backwaters are formed (and turn into the *beels* so beloved of wallowing rhinos), and fresh islands of silt are deposited and gradually vegetated. But most of this new land is beyond the park's northern boundary, and room for the rhinos is being squeezed.

Yet while living with the Brahmaputra makes life difficult, it also makes Kaziranga what it is. The river is the engine that drives the ecosystem. Each year it turns grasslands into breeding grounds for millions of fish that later swim downstream to become food for the rural poor of Bangladesh. It replenishes the soil and prevents trees from becoming established and gradually turning the wetlands into woodlands. Kaziranga is now one of the last naturally functioning floodplain systems in Asia, where the rise and fall of a great river unchecked by human interference shapes and reshapes the entire landscape. And as such, the park has become irreplaceable for the once-widespread wildlife that's evolved to live in this dynamic system.

There are many more treasures here besides the rhinos. Kaziranga is home to over 1,000 wild elephants and half of the world's wild water buffalo. No fewer than 500 species of birds have been recorded—about the same as in the whole of France—including global rarities like Bengal floricans (small bustards), greater adjutants (large storks), and spot-billed pelicans. Gibbons sing in the patches of forest, and the Kaziranga stretch of the Brahmaptura is one of the last strongholds of the *xihu*, or Ganges river dolphin, a blind and bizarre-looking beast that for some reason catches fish in the murky waters of the river by swimming on its side. Even the swamp deer that skittered around our dawn elephant party belong to an otherwise near-extinct subspecies. And the population of tigers that keeps them and thousands of diminutive hog deer on their hooves is the densest in the world.

For me, though, almost as marvelous as the creatures themselves is the paradox of their persistence. All these species have been allowed to flourish in a region with a long history of bloody civil unrest, and in a park that's surrounded by more than 70,000 people. Where household incomes are often less than US$10 a month—and where a rhino poacher can earn many times more in a night than a farm laborer can get in a year. So why on earth has this success story been possible?

Tough Love

There are two overwhelming reasons. One is the dedication and bravery of hundreds of conservation professionals. They repair roads and camps; keep Kaziranga's jeeps, boats, and working elephants in running order; carry out managed burns to hold the growth of grassland trees in check; and undertake annual censuses of the park's residents. But above all, they patrol for poachers.

Apart from a brief upsurge in the 1960s, poaching rhinos for their horns was a low-level problem over the first 75 years of Kaziranga's existence. Rhino numbers increased steadily, reaching around 1,000 animals. But then demand for powdered horn in eastern Asia escalated, and everything changed. In 1981 two dozen Kaziranga rhinos were killed and their horns hacked off. Three years later, the annual toll had almost doubled. The poachers used a range of methods. They shot animals. They dug deep pits along animal paths and lined them with sharpened bamboo stakes that impaled the rhinos that fell into them. And in places where electricity cables cross the park, they threw wires over the power lines and back down onto rhino trails to electrocute passing animals. Laokhowa Wildlife Sanctuary—a smaller reserve nearby caught up in a separatist insurgency—lost almost all of its rhinos to poachers. Kaziranga had many more animals but was by now losing nearly 50 a year. Clearly something had to be done.

The man credited with stemming the tide is Shri Paramananda Lahan. Steely, hardworking, and utterly uncompromising in his commitment to Kaziranga's rhinos, Lahan quickly made anti-poaching patrols the park's top priority. Despite now being long retired and evidently unwell, he kindly agrees to meet with me at his house. He has a kind face and gentle voice but a vice-like handshake. "It was tough at first," he recalls. "There were very few roads, so we moved around on elephant and on foot." Lahan led from the front, tracking poachers, confronting them, exchanging fire: "I was with the guards all the time. We tried to intensify patrols as much as possible. There were only 70 or 80 guards for the whole park, and it took four and a half years, but eventually the area came under control." Rhino killings leveled off, then declined. The population began increasing once more.

Lahan's legacy—carried forward by a succession of equally impressive leaders—is a sophisticated, almost paramilitary anti-poaching operation. There are now nearly 500 front-line staff, stationed across 120 camps scattered through the park. The guards are linked to headquarters by radio, wear army-style uniforms, and for the most part carry guns—albeit older and less powerful ones than their adversaries. Each night they go on patrol in their sector, walking silently, searching for intruders and trying to avoid tigers, bears, and elephants. Encounters with poachers frequently end in gun battles, with both sides shooting to kill. Four guards have died; twice as many have been seriously injured. Yet for the most part, the poachers come off worse. More than 600 have been arrested; over 100 have been killed.

In reality it's a low-level war. Rhino poaching rates are generally down—in the single figures every year from 1998 to the end of 2006—but it's an unending fight. "Kaziranga has to be continuously protected," warns Shri Lahan. "If we stop, even though it looks like there are plenty of rhinos, they won't last long." Newcomers have brought new approaches to the anti-poaching work. They've built networks of informants in the villages. If someone suddenly starts buying larger bags of rice in the market, questions are asked. Other groups try to tackle the real villains in all this—the traders who commission the poaching gangs and move the horns into Burma or Bhutan, and from there to China and Hong Kong. But still most of the work to save the rhinos is done the hard way—on foot, at night, and deep in the heart of the park.

The Unicorn Defenders

We clamber into the small Suzuki jeep and set off down the track. To find out more about the life of a rhino guard, I've been given permission to travel up to the front line to a camp called Arimora, far beyond the tourist zone and right on the park's northern boundary. My guide for the morning is Dharanidhar ("D. D.") Boro, the officer in charge of the 240 men protecting the so-called Central Range of Kaziranga. Mr. Boro is an interesting man.

He's a Bodo,[3] from one of the main ethnic groups in Assam. Starting as a village cowherd, he's worked his way up through the Assam Forest Department, en route collecting a law degree and a cabinet full of awards for his contributions to conservation. Round-faced and with a bright-eyed intensity I come to recognize in the many of the leaders I meet on my journey, Mr. Boro combines a love of flute playing and poetry with the same fearsome determination to conserve rhinos as Shri Lahan. He's opened up new fronts in the fight against poachers—getting involved in the lives of the villagers around Kaziranga; trying to persuade them the park is their property, not his; even recruiting a network of ex-poachers as informants. As he explains, "If you have a pin in your foot, you use another pin to get it out." But he has Lahan's reputation for toughness too. One of his commendations reads, "For more than 100 encounters with well-armed poachers, in which deaths have been common." The amiable-looking man alongside me in the car has shot a lot of people.

From Central Range HQ it's a short drive past the lush, even bushes of a well-manicured tea plantation and across the NH-37 highway into the park itself. Once inside, we drive north along a causeway, great stands of elephant grass stretching away on either side and the faintest snow-clad peaks

---

3. Somewhat confusingly, Bodo is pronounced "Boro." This presumably explains why Boro is one of the commonest Bodo surnames.

of the Tibetan Himalayas just visible in the farthest distance. Elephant turds are scattered along the road like giant Christmas puddings. The next *beel* reveals the wild elephants themselves, loafing massively around the edge. A small group of wild buffalo rest in the foreground, their huge sickle-shaped horns reflecting in the water.

The signs of rhinos are all around. Their well-trodden trails (known locally as *dandis*) lead off from either side of the road. Every few hundred meters we pass an enormous roadside latrine—the accumulation of months of nightly dunging by the local rhinoceros. At one, a copper-red and green jungle fowl—svelte progenitor of the domestic chicken—scratches for insects, then dashes for cover as we get near. From the dungpiles and trails it's clear that Kaziranga's rhinos are creatures of habit: easy targets for determined poachers.

We cross the Diphalu River by hauling the car along a rope ferry, and soon we're there. Arimora. A simple hut raised 3 meters up on concrete stilts to withstand the floodwaters from the Brahmaputra, which runs right alongside: immense, brown, and deceptively sluggish. This is conservation's front line—1,600 kilometers upstream to the Tibetan glaciers that feed the river, another 1,300 kilometers downstream to where its delta floods the great mangrove swamps of the Sundarbans—yet only a few minutes by boat from the islands where the poaching gangs await their chance. Arimora and its ilk are where the survival or otherwise of the Indian rhino is being decided.

I meet more of its guardians—seven middle-aged men in khaki, with old but well-maintained .315 rifles slung over their shoulders. They explain that they spend a month at a time out at this small, spartan camp, then get a couple of days at home with their families. By local standards the job is quite well paid—permanently employed guards get around 8,000 rupees (about US$160) per month—but it can be lonely, and it's frequently dangerous. For company, each camp has a cat; to help come to terms with the risks, each has its own Hindu shrine as well.

Praying for good luck seems like a sensible idea. The wildlife the guards are trying to protect pose enough of a hazard: Mr. Boro recounts grisly stories of tiger attacks and marauding sloth bears. But the guards' real worry is being ambushed by poachers. It's a nightly danger. They tell me that they patrol their area every evening, usually in a group of three. They walk for 10 or 15 kilometers along the roads and *dandis*. To avoid detection by poachers they don't use torches, and they communicate with other guards using a private system of animal calls—mimicking frog croaks and bird whistles. They find poachers every few weeks, their presence typically betrayed because, unlike the guards, they don't wear boots and so leave different tracks. Then the guards patrol until they find them.

The ensuing gunfights seem chaotic affairs. Bhimlal Saikia—recruited, like most of the guards, from a nearby village—recalls a September night 15 years ago: "It was one o'clock, and there was a full moon. Poachers had tried to kill a rhino. They missed, but we heard the shot, and several groups of guards came into our area. Then we waited for a long time. At 4:45 the poachers suddenly moved toward our group. There were six of them and only three of us. They were wearing khaki, so we thought they were guards and shouted, 'stop.' They didn't, so we knew who they were. They kept coming. One of them had a .303 rifle, so I shot him. He died straightaway. The others ran away." Were there repercussions for him, I ask? "The next day the range officer came over with the police. They found the poacher's rifle and ammunition. I didn't get any problems, but they moved me to another camp so other poachers couldn't come after me."

Each man around me tells a similar story of dark encounters, of confusion punctuated with gunshots, of people dying over rhinos. I'm struck by how matter-of-fact, how unembellished their accounts are. No euphemisms, no tall tales. This is simply what these men do in the name of conservation. And right now, things are getting worse. Poaching is on the rise again in Kaziranga: more than 20 rhinos were killed in the year of my visit—the worst for over a decade. The guards go patrolling again in a few hours. I hope there are no poachers out there tonight.

I wish the guards luck and say goodbye. On the way home Mr. Boro relaxes a little more. He talks about how proud he is of his men. He tells me his own stories from the field and about the fragility of the Kaziranga ecosystem. "Every creature has interactions with another. If we lose one finger, our hand won't work as well. So it is with species." We cross back over the Diphalu on the little ferry and then drive through a narrow cane break—a jungle of sprawling, spine-coated rattans climbing through the lush bankside vegetation. Just as we get out into the open again, the jeep brakes sharply. My guide has spotted something astonishing 200 meters to our left. Lifting my binoculars, I finally see it: crouching by the sunlit edge of a small pool is the most sought-after animal in India. A tiger, glowing amber against the green. I gasp. I can't believe my luck. Tigers are relatively common in Kaziranga but are rarely seen, even when the vegetation has been burned back. At this time of year they're scarcely encountered at all. The huge cat sits up—it's seen us too—walks three paces, and is gone.

Everyone in the jeep is delighted for me. We drive on, big smiles all round, and the talk turns to tales of tigers. Ones that were seen more than fleetingly, ones seen rather too close for comfort, ones almost walked into while out on patrol. The female that became a cattle killer and recently attacked the park staff who were trying to . . . but Mr. Boro is interrupted mid-sentence by his mobile phone. It's his director, and it's bad news. Informants have reported that a group of poachers has entered the park from the west and is heading toward the Central Range. The atmosphere in the car changes abruptly. Stories of past adventures stop immediately, and we rush back to headquarters.

Golap's Story

The struggle between Kaziranga's rhino poachers and its rhino defenders is deadly serious, but it's also deeply human. I get an extraordinary and unexpected opportunity to understand what that means when I meet up with a local journalist and rhino activist named Uttam Saikia. Slight, charming, and with a fiercely intelligent gaze, Uttam divides his time between writing for local news agencies and running Bhumi, a small, Kaziranga-focused NGO that works to raise awareness of the area and ease conflicts between the park and neighboring villagers. In the last few months his investigative talents have helped him get close to several of the poachers responsible for the latest upsurge in rhino killings, but he's not simply after tracking them down—he wants to help them get out of poaching forever.

We talk about his work, about Bhumi's radical ideas for rehabilitating poachers and setting them up with a fresh start. And then Uttam comes up with an idea that changes everything. One of the most-wanted men in this part of Assam—who is linked to several rhino deaths this year alone and on the run from the police and the Forest Department—is close by and apparently wants to talk. Uttam phones him, I hastily rearrange other plans, and we fix a time the following day. As a lifelong conservationist (and someone who, as a child, went on sponsored Save the Rhino walks), I'm about to come face to face with the enemy.

The poacher who wants to talk is named Golap Patgiri. When we arrive at the meeting place—a poorly lit café with no other customers but plenty of flies—he's already there, sitting at a rickety oilskin-topped table with his back to the street. Golap is tall and thin, wearing an old but carefully ironed shirt and trousers; his careworn face seems much older than his 41 years. He also looks completely terrified. Uttam introduces me, but Golap scarcely looks up. We sip at scalding-hot chai and wait in silence. I worry Golap has changed his mind, but then he says a few, hushed words to Uttam. It turns out he's worried the waiter is listening in. He still wants to talk, but we need to go elsewhere. He agrees to get into our jeep and drive to his village.

We travel along a bumpy cycle track perched between rice paddies. Simple bamboo homes stand on stilts by the edge of the water, and small mud-caked children work the flooded fields, trying to scoop up tiny fish using what appear to be large, open-weave trays. I start to understand what trying to live on US$10 a month must actually be like. The children shriek at each other then grin broadly and wave at us. But life is tough. We pass a village right next to the park where a rhino was killed in broad daylight just last month. So far this year seven have been poached around here.

Golap's village is nearby. We park, take off our shoes, and climb the ladder into his house. His three-year-old son, Bharat, rushes up to hug his dad, squeezing tight around his legs. Uttam and I are shown to the only mat and offered rice wine and something that tastes like rice pudding. Golap sits on the bamboo-slatted floor with Bharat curled up on his lap. More at ease, he at last starts his story—softly, almost conspiratorially, all the while looking down at the floor.

"I used to work in the park—for nearly 10 years I worked as a 'casual' guard. This meant we did the same work as the permanent guards, but we only got paid 1,500 rupees [around US$30] a month. We wanted to be made permanent, but they never had the money. Then in April 2000, they said they couldn't afford us anymore. So I had no money and a family to support. It was a very difficult time." He explains that social convention meant he felt unable to take menial work: "I'd had a good job. Everyone knew I was a guard for 10 years. So how am I supposed to start plowing or laboring? It would bring me too much shame. I thought for a long time. And then I started to think about going to kill rhinos."

Golap tells me he put his plan into action in late 2005. He traveled to Dimapur, just over the border into the neighboring (and distinctly less lawful) frontier state of Nagaland. There he met two men from Mizoram[4] and struck a deal to escort them into Kaziranga. "On 10 November I brought them down from Dimapur. I didn't come through the village. I didn't want my family to know. The men brought a .303 rifle and some food, and we spent the night on the edge of the park." I ask Golap if he was frightened,

---

4. Another of Assam's unruly and mountainous neighbors, which together with Nagaland and Manipur, seem to harbour most of the rhino poachers and horn traders operating in India's northeast.

but he says, "No—our gun had a longer range than the guards' .315s. If they came in front of us, we would kill them, otherwise we would die. As we entered the park we thought we were either going to die or become rich."

At this point Bharat wriggles on his dad's lap, eager for attention. Golap strokes his hair gently. The little boy settles. "At seven the next morning we killed a rhino. We cut off its horn. Then we spent all day hiding in the jungle, because we couldn't come out. Finally, at eleven p.m. we came out and took the horn straight back to Dimapur. From there we went to Imphal [the capital of Manipur] and sold the horn—but it was small, so we only got 6 lakhs [600,000 rupees, or around US$1,200] for it. My share was 2 lakhs."

Uttam tells me Golap's account is pretty typical. Most poaching is done by gangs of three—a shooter, a bearer, and a guide. While the first two are often Nagas or other hill people with links to the traders, they usually recruit disaffected locals to help them find their way around the park—marginalized Karbi or Mishing tribespeople, migrant workers laid off from the tea plantations, or (as with Golap), experienced former guards knowledgeable about how anti-poaching patrols work. In Golap's case it was a profitable arrangement. The gang struck again, three months later, but this time weren't so lucky.

"The next time, I went back to Dimapur, and brought them down here again, near the village. We entered the park in the nighttime but we didn't find a rhino until the next day. We shot it in the head." Golap goes on to describe how he took the horn, but on their way out his accomplices got caught and beaten by villagers patrolling for cattle thieves. "I got away, but the villagers called the Forest Department people. The men from Mizoram told them my name." After a month the authorities found Golap and arrested him. He had hidden the horn and wouldn't tell them where. "Everyone was angry and everyone beat me," he says. "I went to jail, but after 25 days I paid bail and came back home. I sold the horn and got 1.6 lakh [around US$320]."

The case has yet to be settled in court, but Golap's exposure as a poacher has already cost him a great deal. After his release, he was pretty much ostracized by his people: "The villagers told one another not to mix with me, not to go to my house. My father said that if you kill a rhino it is very, very bad." Most of Golap's neighbors agreed and forced him to move away from the village, his wife, and his children. Golap stops talking, and draws his arm close around his son. Even now, this is a rare trip home.

After a pause, Uttam quietly takes up the thread. It seems that things didn't end there. Despite the beatings, despite the reaction from his village, Golap carried on poaching and led gangs into Kaziranga three more times.

Each time they killed a rhino, and each time they managed to evade the guards. But news of Golap's involvement got out, and he became a wanted man—unable to go home and unable to go near the park for fear of being shot on sight.

With the Forest Department and police closing in, Golap turned to Uttam's Bhumi organization for help. He says that despite the poaching he has very little money: "Everything I had has gone to lawyers and to the police." Golap tells Uttam he now wants to stop poaching for good, rejoin his village, and become an informant. "I want to become a good man," he says. Uttam believes him, and through Bhumi is trying to raise enough money to set him up in a small business. There's no shop in the village, and Golap wants to become a shopkeeper. Uttam believes building alternative livelihoods like this is essential. "I want to give Golap the opportunity to restart his life—for himself and so that his children respect him. But it's really up to him."

First, though, Uttam is trying to negotiate an agreement between Golap and the Forest Department: the authorities drop all the cases against him, and in exchange one of the most wanted rhino killers vows to go straight and never enter the park again. It's a gamble all around, but unless both parties take the risk, more deaths seem inevitable: most likely more rhinos, possibly a guard, quite probably Golap. As a sign of good faith, Uttam asks Golap to give up a handmade gun he has stashed away. Golap says he's keen. Just before we leave, he talks about surrendering in the next few days. Of all things, he's in a particular hurry because Bhumi is organizing a 220-kilometer cycle ride next month to raise awareness about Assam wildlife, and he wants to take part: poacher turned proselytizer. But Golap is also tired of being on the run; he wants to go back home and be welcome there once more.

"Father, Please Go Away"

The villagers' reaction to Golap brings home to me the second big reason why rhinos are still wandering around Kaziranga. Rhinos remain (as do elephants, buffalos, and tigers) because many Indians want them. You don't have to spend long in Assam to realize that local people are tremendously proud of Kaziranga in general and its rhinos in particular. The rhino is the state animal, and a stylized rhino is the logo of choice for almost all enterprises Assamese, from bus companies and tea estates to the state oil company and the local squadron of the Indian Air Force. Pride in rhinos brings large numbers of Assamese tourists to Kaziranga. It makes rhinos and what happens to them a frequent talking point for journalists and politicians. It even means that the Forest Department has received the occasional helping hand from the United Liberation Front of Assam,[5] one of the main militant groups fighting the Indian government for an independent Assam.

More broadly, local people's respect for big and distinctly dangerous creatures is manifest in their quite extraordinary tolerance of the damage that they cause. Across much of India, large mammals persist in areas of dense human settlement to a degree that is unparalleled just about anywhere else. There are over 1 billion Indians, living at an average density of over 300 people per square kilometer—three times the population density of the European Union, and 10 times that of the United States—yet somehow they manage to coexist with sizeable populations of elephants, leopards, and wolves.

The human costs of living so closely with large animals are very significant. These wild creatures variously eat crops, damage fields, flatten homes, attack livestock, and even injure and kill people. In an average year in the state of Madhya Pradesh alone, such human-wildlife conflict costs poor rural communities over US$170 million; over 600 people are injured by tigers, leopards, or bears, and of these, 30 or so die. Across India as a whole, elephants kill around 300 people annually. In the most shockingly affected area—the mangroves of the Sundarbans, along India's border with Bangladesh—tigers routinely kill up to 100 people each year. And then eat them.

Even more striking than this dreadful toll, however, is the Indian public's general acceptance of it. Not long ago in Europe, the first wild bear seen in Germany since 1835 lasted only a few weeks before accusations of killing sheep (not to mention raiding a rabbit hutch) led to it being shot dead. Yet in India, even when they kill humans, rogue wild animals are usually spared; on average only one elephant is killed for every ten people killed by elephants.

The reasons? Everyone I've asked attributes this extraordinary tolerance to firmly held religious beliefs regarding the rights of other creatures. While not many people go as far as those Jains who sweep the floor in front

5. ULFA is "a revolutionary political organization engaged in a liberation struggle against . . . India" or a banned terrorist organization, depending on your viewpoint (see en.wikipedia.org/wiki/United_Liberation_Front_of_Asom, accessed November 21, 2011). ULFA allegedly went out if its way to assassinate several rhino poaching ringleaders in the early 1990s. As one former Forest Department employee told me, "Why not? It makes good sense. ULFA is an Assamese organization, and anyone who loves Assam loves rhinos."

of them to avoid treading on insects, for most Hindus and Buddhists (which in practice means most Indians), all animals are divine. The great god Ganesh has an elephant head. Other animals too are the living incarnations of deities. Deep-rooted fatalism is also important. A common remark when people are killed by animals translates roughly as, "This had to happen. This death was in his destiny."

The Assamese pay a particularly high price for their acceptance of their fellow creatures. The state has one-fifth of all of India's elephants in one-fortieth of its area. On average these kill 60 men, women, and children a year—and as conflicts grow, thanks to remaining habitat patches shrinking, and corridors between them being converted to farmland, the death rate is rising. Crop raiding is less devastating but much more common. Around Kaziranga, villagers report that over a quarter of their rice harvest is lost each year to wildlife. Elephants are the main culprits, but wild boar, deer, rhinos, wild buffalo, monkeys, and even parakeets raid the fields, leaving poor farmers poorer still.

They try their best to defend their crop. The flat paddy landscape is punctuated with fragile, stilt-legged lookout posts called *tangis*, from where dusk-till-dawn sentries can give early warnings of approaching animals. Villagers then try to chase the would-be crop raiders back with burning torches, drums, and firecrackers. One of Golap's neighbors explains: "We fight every night to protect our crops from wild animals. We use fire and spears and drumming to chase them back to the park. We never try to kill them—we just chase them back." They sometimes succeed, but often don't. "By itself, one elephant can eat two or three quintals[6] in a night. Sometimes 50 or 100 elephants come and destroy the whole area. Our houses can get broken too. And sometimes even tigers come and injure cattle and people."

On the way back from the village, my driver offers to take me to meet one of his neighbors so I can see the impact of cropraiding firsthand. It's late in the afternoon, and the lush greens of the paddies and the park floodplains beyond are intensifying in the yellowing sunlight. I start at a sudden gunshot far in the distance. I scan the horizon and spot a rhino a couple of kilometers off, making its way from the cover of the park into a field. It's the time of day for crop raiding to begin, and ripening rice grains are very tempting. Two more volleys, and then I realize they're not gunshots, but firecrackers. The rhino stops sharply, looks around, and trots back toward

---

6. A quintal is a unit of mass which helpfully means different things in different places; in India it is equivalent to 100 kilograms.

the park—one unwelcome visitor successfully dissuaded. But many more are not.

We walk along the raised mud walls between paddies to the edge of a field flattened by elephants 15 days before. Matiram Phukon, a slight 56-year-old man wearing a jumper and *dhoti*, gestures at his ruined crop. "They came in the night—six or seven of them. We tried to drive them back to the park with fire and drumming, but they are very dangerous. They ate only some of the rice, but they trampled a lot more." Altogether he reckons he lost around a quintal that night—one-thirtieth of his entire crop. And that was before the rice had properly seeded. Now the grains are ripening, and Matiram's expecting the elephants to return soon.

I ask him about compensation. It is supposed to be paid, but it's very little and it arrives late—if at all. Most villagers have given up applying. "I told the Forest Department, but they didn't come. Nobody came." One study conducted around the area suggests Matiram's experience is typical: local people pay a punishing price for living near the park. I look around at the destruction and try to comprehend what losing the return on months of hard work, time after time, must be like for someone struggling to feed his family. What does he feel about elephants, about Kaziranga, when he has to live next to them?

Matiram breaks into a broad, gap-toothed smile, and then says something that staggers me. "We don't want to hurt the elephant. We also care for the elephant. If he comes and eats our crops, we feel angry. But not all the elephants or rhinos are doing that, and we're only angry with the ones that come to eat. The rest of the time, we love Kaziranga." My translator, Polasz, sensing my astonishment, tells me what happens in his own village: "I have a rice field too, and after planting I lay a coconut, some gram, a betel nut, and an oil lamp onto some banana leaves and pray to Ganesh that he will not come and destroy our crops. We believe it helps; one time the elephants came across the edge of my field but didn't destroy it. A lot of people do this thing." And if they destroy the crops anyway? "Then we say to them, 'Father, please go away.'"

We leave the field just as the sun sinks below the skyline, turning the clouds orange and pink and purple. Matiram walks toward his hut, and I climb into the car for the short ride back to my comfortable lodge—shocked, amazed, and extraordinarily humbled. I may send a few readily afforded pounds to the conservation organizations that help support parks like Kaziranga, but when I compare my contribution to that of Matiram and his family and their neighbors, I've no doubt who the real conservationists are.

Which Way Conservation?

Mr. Boro's men risking their necks nightly for US$5 a day; Golap on the run and ostracized from his community for trying to make a bit more than a farmer's pittance; Matiram matter-of-factly ruing his crushed crops but forgiving the perpetrators: Kaziranga embodies conservation at the sharp end. It forces us—by whom I mean wildlife-loving, mostly Western, mostly comfortably off conservationists—to confront some stark realities.

Shri Lahan's ruthlessly enforced protection regime for Kaziranga's rhinos makes us wince: conservation's not supposed to be about shooting poor people, whatever they've done. Despite being responsible for the deaths of five rhinos and being prepared to kill his ex-colleagues, Golap Patgiri doesn't strike me as an evil man (and I'm no longer sure he's the real enemy). Under the same difficult circumstances how many of us would have done the same thing? And the price that Matiram and his neighbors pay year in and year out because they're not as intolerant as the rest of us when it comes to living alongside damaging and dangerous creatures: for all our rhetoric about the need to share the planet with other species, which of us would willingly put ourselves in Matiram's shoes? We—the mostly well-off conservationists and nature lovers who want to live in a world with rhinos and tigers and elephants—are free riders in the system, and we owe Matiram Phukon and millions like him a huge debt for their forbearance. But the tolerance of those in poor farming communities across India (and indeed, large parts of Africa and Latin America) can no longer be taken for granted, and many are beginning to argue that more of us should be footing a fairer share of the bill.

Conservation is responding. Concerns about its human costs are leading to changes in the way it is being practiced on the ground in Assam and beyond. Mr. Boro's ideas about engaging with local people are becoming the norm; Uttam Saikia's more radical proposals to provide alternative routes out of poverty are becoming widespread. In some places those paying the up-front costs of conservation are beginning to receive payments for continuing to do so from distant beneficiaries.

I'll find many more examples of these sorts of innovative approaches to conservation through the rest of this book—places where conservation is working because it directly addresses poverty, improves local people's everyday lives, or even makes good business sense. Linking conservation to communities is the way of the future. And Kaziranga's future, like that of many other wonder-filled places, will come to depend on how far it contributes to the well-being of the ordinary people that live around it.

But for all that the old-style, fortress-and-fines conservation may make us feel uncomfortable, and as much as many conservationists may want to move away from it as swiftly as possible, perhaps it has its place. Where would Kaziranga's rhinos be now without the guns and the guards who have protected them by brute force and bravery over the past quarter-century? Many would argue nowhere—they would have gone the way of just about every rhino population without an armed guard. And what then would we prefer had happened? Guns and rhinos, and a nagging sense that there ought to be a better way, or bottom-up, community-centered conservation and very probably no rhinos at all? Kaziranga and the prospects for its future force us to face a fundamental question. What exactly does conservation want to do, and how does it best meet its own legitimate objectives in the face of overwhelming human poverty?

Another Century of Rhinos?

So what does the future look like for Kaziranga? Will its 2,000-odd rhinos remain, beyond the 2000s? Will my great-grandchildren and yours live in a world that still has space for unicorns, and will Golap's and Matiram's great-grandchildren see them when they look beyond their fields? Over the short- to medium-term, the future of Kaziranga and its spectacular inhabitants will depend, as in the past, on keeping poaching under control and maintaining the backing of the local community.

The bravery of the park's rangers and the commitment of their leaders will continue to be vital, but other anti-poaching initiatives will probably become increasingly important too. If it succeeds, tracking down and prosecuting the traders and breaking up the supply chains that smuggle poached horns into China would yield disproportionate dividends. Efforts by Bhumi and others to rehabilitate key poachers may also prove very worthwhile. And strategic reintroductions now underway to put rhinos back into other reserves where they've been wiped out should help spread risks.

What about prospects for the continued tolerance of wildlife by Assam's rural poor? All the signs are that human-wildlife conflict is set to worsen. As human populations grow and expand, animals are inevitably coming into ever-more frequent contact with farmers and their fields. And as traditional ways of life become exposed to Western values, commercialization is beginning to take hold and attitudes are starting to harden. To counter this, ways of spreading the economic benefits of the park among its neighbors are being developed: some enlightened tourist operations employ only local people, visitors can spend money on community-made handicrafts, and so on. But realistically, there are limits. Kaziranga already 70,000 people living on its doorstep. The park would have to be generating revenues of many millions of dollars each year to have significant impacts on most of its neighbors' household budgets—and that seems a very distant prospect.

There are larger-scale issues, too, starting to loom over the horizon. Most of them are about the need to conserve animals and plants and the places they live, not as if they were static entities, but to consider the dynamic processes that maintain them, have shaped their evolution, and are now coming under threat. The importance of processes is something that dawned on most conservation scientists (me included) about 15 years ago, but it's something that real-world practitioners have been wrestling with for much longer.

In Kaziranga, the dominant process to be accommodated is the flooding and shifting of what has euphemistically been labeled "the naughtiest river in India"—it is vital to the park's existence but challenging as well. One great challenge the Brahmaputra's bad behavior presents is the need to replace what it's swept away with newly formed land. Progress is slowly being made in expanding the park's boundaries northward to achieve this. A second set of problems arises from the need for animals to escape particularly serious flood events by moving to higher ground. In the past, they moved south onto the forested slopes of the Karbi Anglong Hills: the uplands and the floodplains formed one ecologically interconnected unit. But the hills lie beyond the park on the wrong side of the busy NH-37 highway. Many animals trying to cross it die in the attempt—buffalo, deer, wild boar, even elephants. Those that make it to the hills—especially the rhinos—risk being killed by hunters. Those that stay behind risk being drowned; some years, floods kill more rhinos than poachers do.

There are some helpful responses to these flood-related problems. Park staff have devised an early warning system so they know in advance if the river is likely to burst its banks. Working with the army, they've built dozens of little plateaus—earthwork arks where wildlife can sit out the worst of the flooding without having to leave the park. Hearty roadside mottoes entreat drivers to drive safely: "No hurry, no worry!," "Care makes accidents rare!," "Speed thrills but kills!," and my particular favorite, "Horn do!" And during peak flooding times, dozens of volunteers descend on the NH-37 to escort slow-speed convoys of lorries through the park.

Taking a wider view, though, it's clear that the long-term future of Kaziranga will depend on thinking beyond its current boundaries. The park will face—indeed, is already facing—new threats, often from far away. Proposed schemes to generate hydroelectricity from one of the world's largest untamed rivers risk fundamentally altering the entire Kaziranga ecosystem. Pollution from upstream agricultural intensification could do much the same. And plans to expand the NH-37 into a six-lane super-highway could, if not redirected, greatly increase the isolation of the park. Politicians who pride themselves on the marvels of Kaziranga need to think through the likely impacts of all of these proposals on its survival. They also need to give serious consideration to the idea of extending protection southward into the Karbi Anglong Hills so that an entire functioning landscape is conserved, rather than one large but increasingly vulnerable fragment.

So is there hope? Keeping well-armed, highly profitable poaching operations from taking over; maintaining good relations with tens of thousands of neighbors who can't afford to carry on losing their crops to wild animals; rerouting major transport proposals; implementing landscape-wide conservation; and (quite literally) making up lost ground: in many ways these problems seem overwhelming. But wind back just over 100 years to a time with fewer than 20 rhinos, no outside interest, zero protection, and lots of trigger-happy autocrats, and prospects must have seemed far bleaker. The story of Kaziranga shows us in spectacular style that the improbable is possible. If the immense dedication of Assam's conservation professionals continues and the extraordinary respect its people have for their fellow creatures endures, then my guess is that there still will be unicorns roaming Kaziranga in the twenty-second century. If we all want them enough, they will be there. And future generations will be the richer for that.

ENDING THE WOODPECKER WARS

A switch in context—from Asian poverty to relative American prosperity—reveals that conservation problems are still essentially about resolving people's aspirations with the requirements of other species but also shows how lateral thinking can help. In this case, lawyers have worked out how to sidestep an unintended but pervasive downside of the U.S. Endangered Species Act so that it now rewards rather than punishes private landowners keen to help nature.

I'm standing propped against the rough bark of a tree, peering through the slanting shafts of evening sunlight for signs of movement. The two round holes on the pine trunks in front are still in sight, but there's no one at home. A group of brown-headed nuthatches—gray-backed and squeaking like children's toys—pass nearby. Delightful to watch as they energetically glean the branches for insects, but not what we've come here to see. The nuthatches move on, and we settle again, straining our ears, and scanning the canopy overhead. The sun slips below the trees, and the wood around us falls briefly silent. And then, just as I start wondering if we are going to be unlucky, the small party of birds we've been waiting for edges into view.

One at first, then a second. Two more, off to the left—four in all. They are dapperly marked with bold white cheek patches and black-and-white-barred backs and scattered across a couple of flat-topped pines. Twittering with one another while they travel, the birds work their way busily up the trunks. I watch them feeding. Bark flakes patter to the floor as the birds flick off loose pieces, then hammer briefly and probe the exposed wood for roaches and beetle grubs. A final session of foraging before bedtime.

Then suddenly one bird flies over to a nest hole, dives in, and is gone. I switch focus to the second cavity and see a sharp bill briefly protruding before it too disappears. Another animal, engrossed in a particularly rewarding crevice, stays out a while longer then looks up and flits behind a tree, presumably into another hole. I search for more movement among the branches, but the last bird has already left.

Party over, my companion turns and smiles, delighted to have shown me the animals she's worked on for more than 20 years. "There ya go," she whispers in a gentle southern accent, "arsey-dubbyas!" No, not grumpy former U.S. presidents, but RCWs—red-cockaded woodpeckers: starling-sized, workmanlike, and, these days, extremely rare.[1]

It's a few months on from my trip to Kaziranga, and I've traveled to the opposite side of the world, longitudinally and in countless other ways, to an ancient stand of longleaf pines in the Sandhills of North Carolina. Conservation battles are fought very differently in the United States than in Assam—in courtrooms and with consultants rather than in nighttime crop raids and shoot-outs. But the underlying struggles still center around reconciling the needs of nature with people's need to make a living and, just as in India, prospects for progress seem to lie in the imaginative use of carrots as well as sticks.

I'm in search of woodpeckers not just because they're amazing in their own right,[2] but because their story illustrates what can happen when well-intended conservation legislation has unintended consequences. Across the country, rigid enforcement of laws designed to ensure protection for endangered species like RCWs meant that discovering them on your property—and worse still, helping them multiply—guaranteed increased restrictions on how you could manage it. For many farmers, ranchers, and foresters this caused immediate economic losses; for others it severely constrained future options. Either way, it was scarcely encouragement to do the right thing.

The red-cockaded woodpecker swiftly found itself at the center of the resulting conflict between landowners and conservationists—and then the target of a bold attempt to resolve it. So I've come here to the longleaf pine forests of the Southeast to find out how the troubles of this particular, rather modest-looking bird have prompted changes in the implementation

1. RCWs are rather oddly named, as they're essentially black and white, with less red than almost any other woodpecker on the continent. A cockade is a small rosette worn on a hat to signify the wearer's political allegiance or rank. Juvenile male RCWs have a small red patch on their foreheads, females have no red at all, and adult males sport a tiny red cockade above each cheek but keep them hidden unless they get excited. And although I was really excited to see them, evidently the feeling was not reciprocated.

2. For instance, woodpecker adaptations for hitting their heads very hard into trees up to 20 times a second include a complex array of muscles and tendons that when contracted work as shock absorbers so the birds don't give themselves brain damage, and reinforced third eyelids that close milliseconds before each strike to stop their eyes from popping out of their sockets.

of the most powerful conservation law in the land, and along the way improved the outlook for dozens of other threatened species.

Caught in the Act

The Endangered Species Act, signed into law by Richard Nixon in December 1973, has for over three decades been the toughest piece of environmental legislation on the U.S. statute book. ESA offenders can be fined up to US$50,000 or sent to prison for as long as a year. Explicitly designed to prevent animals and plants from going extinct as a consequence of unfettered economic growth and development, the ESA applies not just to publicly owned land (where it requires the authorities to increase or at the very least maintain populations), but to private property too. Here the regulations are less stringent, though (for animals at least) they still outlaw any activities that result in the so-called take of listed species—where "take" is defined as harm, not just to the creatures themselves, but also to the habitat they're living in.[3]

To receive ESA protection a species has to be formally listed as either endangered (because it's likely to go extinct unless the situation improves) or threatened (because it's likely to become endangered). The magnitude of the extinction crisis across the States means that the agencies responsible for implementing the ESA—the National Marine Fisheries Service at sea and the U.S. Fish and Wildlife Service everywhere else—have plenty to do. So far over 1,300 species have been listed, though the lack of knowledge about many of the most species-rich groups (like insects and fungi) and the paperwork needed to formally list a species are together so overwhelming that some authors reckon the true number of imperiled species is more than 10 times this figure.

ESA protection has led to some impressive successes. The bald eagle—symbol of the nation, but by the 1960s prophetically close to nationwide extinction everywhere outside of Alaska—increased more than twentyfold before being triumphantly delisted in 2007. Whooping crane numbers grew eightfold under the act, peregrine falcons more than quintupled, and

3. This is much stronger than the protection afforded threatened species in many other countries. In the UK, for example, species listed under the Wildlife and Countryside Act are only protected from direct harm, and (for some species) from damage to their breeding sites and resting places; foraging areas and other habitat patches that a listed species depends on are only protected if they also qualify for separate, area-based designation (such as meeting the criteria for a Site of Special Scientific Interest).

the population of grizzlies roaming around Yellowstone doubled. California condors and black-footed ferrets, both of which died out in the wild two decades ago, are now back in significant numbers. Even slowly reproducing gray whales have more than doubled and, like the grizzlies and the peregrines, have been taken off the list.

Not surprisingly, however, given its scope and strength, the ESA has plenty of opponents. Some ranchers and farmers, many mining and logging companies, parts of the oil and gas industries, and elements of the religious and political right—often combined under the loose umbrella of the self-styled "Wise Use" movement—have all fought hard against the restrictions the act places (in their view, unfairly) on their freedom to extract resources from the land. Tensions reached international prominence in the early 1990s, when listing of the northern spotted owl led to a court order banning logging on federal land throughout the bird's range in the Pacific Northwest. Loggers claimed the ban would lead to tens of thousands of job losses, a local restaurant offered "spotted owl soup" (actually chicken and dumplings), and bumper stickers appeared saying "Shoot an Owl—Save a Logger." In the end the Clinton administration brought in a compromise agreement that slowed but didn't end logging—yet thousands of jobs were still lost (more due to automation and the dwindling supply of trees than the ESA) and, with a new threat in the form of larger and more aggressive barred owls taking over their habitat, spotted owl numbers continued to tumble.

The ESA has been under fire more or less continuously ever since. Each year throws up new proposals on Capitol Hill to amend or even ax the act. Under President George W. Bush, constraints on the way officials in the Department of the Interior could operate greatly slowed the process of adding new species to the list—down to an average of only 8 listings a year, compared with 30 to 60 under every president from Jimmy Carter onward. And in one of his final acts as president, in December 2008 Bush relaxed ESA rules to allow federal agencies to permit mining, drilling, and construction on public land without first consulting independent scientists.[4]

Critics of the act say that it simply hasn't worked. Very few species have fared as well as the bald eagle and recovered to the point that they've been delisted—just 22, in fact, as of July 2009. The act's defenders, on the other hand, say this is missing the point. At the time they were listed, all 1,300 or so species were judged to be on the way to disappearing completely. But since listing, only two have actually gone extinct. Completely reversing the

4. . . . . A decision that was overturned four months later by President Barack Obama.

fortunes of the remainder to the stage where they can safely be delisted is bound to take a long time. Yet according to the biannual status updates the Fish and Wildlife Service submits to Congress, populations of around half of all ESA-listed species are now improving or are stable.

Drilling down into the figures, though, reveals a more complicated story with widely varying success rates. Improvements are more common among mammals, birds and fish than among lower-profile creatures like plants or invertebrates. Unsurprisingly, success rates are also higher for species that have more money spent on their conservation, and that have been listed and protected for longer. Most strikingly, there's a substantial difference depending on where threatened animals or plants happen to live. For those found entirely on federal land, almost 60 percent are stable or recovering—but among those living only on private property, the figure plummets to just over 20 percent.[5] Because almost three-quarters of the entire land area of the lower 48 states is in private hands, this divergence in the fate of endangered species has become a serious cause for concern. And it's this major discrepancy in the fate of endangered species on federal versus private lands that the case of the red-cockaded woodpecker has highlighted.

The crux of the problem is that, until recently, private landowners saw the ESA as all stick and no carrot. If the Fish and Wildlife Service knew about an endangered animal species on your land, you could do nothing that risked "taking" (i.e., harming) the animals or indeed any habitat they occupied—even if the damage was accidental. For many landowners, this restriction had serious economic consequences. They were no longer allowed to log their forest, put cattle out on the range, or even repair their farm pond. And there's typically been no compensation available; just as in Assam, private individuals and businesses ended up paying a disproportionate price for conservation that society as a whole has wanted. As a result, landowners have had a strong incentive to do everything possible to ensure they don't have any listed species on their property—and so, ironically, endangered species have lost out too. Fifteen-love to the law of unintended consequences.

"Shoot, shovel, and shut up" (or, "SSS") is the most widely quoted maxim. Taken to the extreme: rid your property of anything endangered before the federal government even knows it's there. Less drastic, but doubtless more common (and entirely legal), is a variant on SSS known as "preemptive harvesting": if your land contains the right habitat for a listed species but

---

5. No figures are available on the status of endangered species confined to the third and smallest category of land—that owned by state or local government.

they're not yet in residence, it's not protected, so you should log it, plow it or drain it before the creature shows up and restricts what you can do. And certainly don't do anything to improve conditions on your property for endangered animals or plants—that'll simply add to your regulatory burden and end up costing you even more. As the National Association of Home Builders ("Reshaping and Enriching Our Communities") has recommended to developers, "The highest level of assurance that a property owner will not face an ESA issue is to maintain the property in a condition such that protected species cannot occupy the property. . . . This is referred to as the 'scorched earth' technique."[6]

There's one more reason why inadvertent incentives for private landowners to do the wrong thing and lack of any inducement to do the right thing have been so damaging: many species are endangered because they require very specific conditions that tend to disappear without active management. Often they depend on habitats that are in the early stages of so-called ecological succession—woodland clearings that haven't yet grown up into dense forests, wetlands that haven't been colonized by trees and started drying out, and so on. In the past, as individual patches thickened up and became less suitable, fires, floods, storms, or maybe large grazing animals would have created new openings elsewhere in the system, which the species could colonize anew.

But as habitats have shrunk and as humans have brought the great forces of nature under control, the persistence of many species in those isolated habitat patches that still remain has increasingly come to depend on active intervention by people—cutting trees, bringing in livestock, keeping scrub encroachment at bay. Without this, the whole of each wetland or woodland or grassland fragment would become mature, and young-habitat specialists would simply die out. So in the stick-only world of the ESA, if a worried landowner didn't want to risk destroying an endangered species' habitat, often all he or she had to do was wait. The habitat would grow up, and the problem would take care of itself.

With such powerful, albeit unintended, incentives for landowners to at best discourage and neglect threatened animals and plants, it's no surprise that endangered species have been faring so poorly on private land. And this matters not just locally but at a national scale, because the United States is, for the most part, privately owned. Consequently, two-thirds of all ESA-listed species are largely dependent on private property, and one-third

6. Craftsman, "Developer's Guide to Endangered Species Regulation," p. 109.

don't occur on federal land at all. So as Aldo Leopold, the great prophet of the U.S. conservation movement, wrote as long ago as 1947, "The only progress that counts is that on the actual landscape of the back forty"—on the privately owned 40-acre parcels of land that cover most of the country.

But by the mid-1990s it was becoming evident that much of that progress was in the wrong direction. To turn things around and become more effective the ESA needed smarter ways of working with private landowners—giving them reasons to do the right thing rather than encouragement to do the wrong thing. And one of the best-known species heading for trouble if this problem was not addressed? The red-cockaded woodpecker, a once-abundant bird by now almost entirely confined to the remnant longleaf pine forests of the American Southeast.

The Forest that Fire Built

Once upon a time, longleaf pines dominated a vast sweep of the bottom right corner of the United States, from Virginia and the Carolinas to Georgia, Alabama, and Mississippi, on as far south as Florida and as far west as Louisiana and Texas. As late as 1700, the land of the longleaf covered a Japan-sized piece of the rolling Atlantic and Gulf coastal plains. Longleaf trees tend to be slender, their leggy, straight stems reaching almost to the canopy before branching out. In favorable conditions they can grow as high as ten-story buildings. But the most striking feature of a longleaf forest is its openness. Rather than forming the dense, dark stands that most of us might think of as forest, longleaf grows as a parkland, with wide gaps between trees. Early European travelers wrote of being able to "ride full gallop twenty or thirty miles on end," "there being no underwood to prevent a horse from galloping freely in every direction."[7]

To quote Larry Earley, longleaf's elegiac biographer, standing in a well-cared-for longleaf tract is to experience "a light and sound show,"[8] with bright sunlit spaces between widely scattered trees, which sough constantly in the breeze like the sound of a distant shore. On the ground, the main plant isn't a tree at all, but the aptly named wiregrass, whose tussocks blanket the floor between the thinly scattered pines. Looking upward, there

7. Earley, Looking for Longleaf, p. 75.

8. Earley, Looking for Longleaf, p. 7.

is far more blue than green, more sun than shade. The eponymous savannah[9] of Georgia's former state capital. A forest you need to wear a hat in.

The ecological force maintaining the openness of the longleaf system is fire. The southeastern United States gets more lightning strikes than anywhere else in the country, especially in spring and summer. Add such frequent ignition to the carpet of shed, 30-centimeter-long needles that give longleaf its name, throw in tinder-like wiregrass and a scattering of fallen, resinous branches, and the result is that left to its own devices, longleaf usually burns every 2 to 5 years. The fires hold in check boisterous hardwood species—bluejack and blackjack oaks, huckleberries, and scrub oaks—striving to fill the gaps between the pine trees. But because the burns are frequent, the fuel supply is limited and so, relatively speaking, the fires are reasonably cool—and the dominant, fire-adapted plant species can take advantage.

Wiregrass burns down but then, freed from its competitors, quickly resprouts. Many other fire-loving specialists (a.k.a. pyrophytes) thrive in the open conditions of the forest floor, making the longleaf system extraordinarily rich in herbaceous plants. Scarlet lilies and candyfloss-colored milkweeds, yellow meadow beauties and indigo birds-eye violets brighten the ground. In wetter corners the mottled throats of carnivorous pitcher plants tempt overinquisitive insects. Likewise the red, trigger-haired jaws of Venus flytraps: the sandy forest soils tend to be low in nutrients so, as at Stoborough Heath, plants able to catch and digest animals fare well in the longleaf.

The pines themselves are insulated by their thick bark from the flames that keep the forest open—so effectively that you can reportedly aim a blowtorch at a chunk and scarcely feel it on the other side. The passing of the fire provides ash-enriched soil into which the longleaf cones can drop their oversize seeds. These quickly germinate and send down long taproots, but above ground they hardly grow at all for several years, instead protecting their sensitive buds in fire-resistant sheaths of green needles. Then at around the age of 7, these seedlings suddenly race upward, putting on a meter or more each year until they're out of reach of all but the hottest fires. So impressive are these adaptations that some ecologists—noting the exceptionally high levels of flammable resins in longleaf wood and

9. Savanna or savannah? It depends on where you were born. To North Americans it's "savannah"; to Brits, Australians, and anglophone Africans it's usually "savanna"; and to Spanish speakers it's *sabana*. But in each case it's a more or less open, grassy woodland with seasonal rainfall and seasonal fire.

needles—have even suggested that rather than responding to fire, longleaf has evolved to promote it: botanical arson as a means of incinerating the would-be competition.

Other species play vital, so-called keystone roles in maintaining the long-leaf system. Gopher tortoises dig large numbers of burrows to evade fires, incidentally providing flame-proof accommodation for snakes, frogs, and countless insects. Highly specialized fungi form intimate associations with the pines—growing tiny filaments called mycorrhizae around the roots of the trees, which soak up sugars from the pines and in exchange nourish them with scarce nutrients gleaned from the surrounding soil. How does a longleaf seedling find a new fungal friend? It seems getting this mutualistic relationship started hinges on the services of a rodent. The southeastern fox squirrel—big and usually black with a handsomely white-tipped ears, nose, and feet—depends at certain times of year on raiding the longleaf cones that no other squirrels in the forest are big enough to break open. The squirrel repays the tree by also feeding on the underground truffles of the mycorrhizal fungus, whose spores it then helps disperse across the forest via its feces. So the squirrel helps the fungus, the fungus helps the longleaf, and the longleaf feeds them both.

RCWs are another keystone of the longleaf system. Unlike almost any other woodpeckers, they don't use dead trees for nest holes—presumably because of the prevalence of fire. Instead, they excavate fresh cavities in living longleaf trunks.[10] To have enough of the inner heartwood to contain a nesting chamber, they generally use trees over 80 years old, and they make their job slightly easier by somehow picking those where redheart fungus has helped soften the wood. Even so it's an extraordinarily time-consuming process, and one that shapes every other aspect of the woodpecker's life. Drilling an upward-sloping tunnel deep into the tree, repeatedly leaving the resulting resin to harden and dry, and then finally boring out a chamber large enough to house 3 or 4 chicks—all this can take 6 years: a real problem for a bird where only 1 percent of fledglings make it to the age of 10.

The solution the woodpeckers have found is to cooperate across the generations. One generation starts the homebuilding, but it's usually their off-spring or grand-offspring that finish it off. The RCWs live in family groups, mostly consisting of the pair themselves plus 1 to 3 male offspring that stay at home helping with construction work and bringing up their siblings. It

10. Although in the absence of a suitable longleaf they will use other species—slash, loblolly, or shortleaf pines.

seems the sons are better off bringing up close relatives and waiting to in-
herit their parents' nest holes (or take over the next-door territory) rather
than dispersing and trying to set up home farther afield.

Not surprisingly, the resulting nest cavities, which tend to be built in
clusters of 3 or 4, are the focus of family life, with the woodpecker group
returning each night to roost in them, even outside the breeding season.
To stop rat snakes from getting in, the birds tear off bark and drill small
wells into the wood around the nest entrance; the resulting streams of off-
white resin create a sticky barrier against predators and make trunks with
occupied cavities look like giant, wax-dripped candles. They also need to
defend their laboriously built homes against squatters. Flying squirrels,
nuthatches, bluebirds, screech owls, wood ducks, and bats, not to mention
other woodpecker species, all use RCW holes and are keen to take them
over if possible.

Until relatively recently, the RCW strategy of hard work and family values paid off. The bird occupied more or less the whole of the longleaf forest and beyond, with an estimated million-plus family groups distributed
from the Pine Barrens of New Jersey to eastern Texas and inland as far as
Tennessee. Thousands of years of use of the forests by native Americans—
collecting firewood, and setting wintertime burns to improve game hunting on the resulting flush of fresh grass—if anything probably helped the
spread of the fire-loving longleaf and its inhabitants. Even early European
settlers largely left the forest to its own devices. But then three things happened. Longleaf became the source of vital raw materials for the world's
navies. Next, mechanization led to its wholesale clearance for timber. And
last, well-intentioned but misguided officials decided that fire was bad for
forests.

The Fall of a Forest

Wander around the piney woods of the Carolinas and it won't be long before you come across a ghost from a now-extinct industry. A gentle dip
in the ground, where a kiln once burned the resin-rich stumps and fallen
branches of longleaf, turning them into tar. Or a "catface"—a great scar
down the side of a trunk, from chest height to near ground level, where
a broad sliver of bark was cut off some 150 years ago and then visited time
and again to collect the accumulated gum, the raw material for distilleries
to separate into turpentine and rosin. The tar was used to protect the miles
of ropes needed by sailing ships or boiled to make pitch for waterproof-

ing their hulls. The turpentine and rosin had hundreds of uses: making industrial solvents and paint, cleaning wounds, relieving corns, treating pleurisy, even helping farmers pluck their poultry and fiddlers keep their bows in trim. Together, tar and turpentine, rosin and pitch were known as naval stores. Through the eighteenth and nineteenth centuries they were as vital to the world's economies as petrochemicals are today. And through most of that time, the world's leading supplier of naval stores was the long-leaf forest of the southeastern states.

The 200-year naval stores bonanza did great damage to the pine forests. A tree can be tapped for resin almost indefinitely, yet turpentiners in particular were needlessly destructive and so had to move on every few years, leaving tens of millions of dead trees in their wake. But the real destruction came not from the demand for tar and turpentine, but with the advent of mechanized logging.

As they exhausted their own forests, northern lumbermen began by the 1850s to set their sights on the great pinelands of the South. Vast tracts were bought for just US$1.25 an acre. Railway companies were given thousands of acres for every mile of track laid—and so built thousands of miles of track. Logging trains and steam-powered skidders and loaders and sawmills moved in, and the world's first industrial-scale clearcutting operation took over.

At its height, the loggers harvested around 7 billion board feet[11] of longleaf a year—roughly enough, by my reckoning, to fill all the containers of a train reaching from Chicago to Los Angeles. Every year. The reason for this mind-boggling harvest? The near-doubling of the U.S. population in just 30 years, coupled with spiraling demand for construction materials. Longleaf became known as "the forest that built America." But the consequences were inevitable. In little more than half a century, nearly all the longleaf was gone, and the giant lumber industry was on the move again, this time to swallow the old-growth forests of the Pacific Northwest.

The catastrophic loss of the southeastern forests spurred government agencies to develop scientific methods for harvesting the remaining forests sustainably—but for the red-cockaded woodpecker, this spelled yet more trouble. Worried about the incendiary conditions left behind by the turpentiners and loggers—dying, resin-coated trees and discarded stumps and branches—Forest Service managers singled out fire as the enemy. Fire

11. A board foot is the standard North American unit for measuring timber, and is the volume of a board 1 foot wide, 1 foot long, and 1 inch thick (or 0.00236 cubic meters). Standard freight containers are 5.94 square meters in cross-section, and I've added 20 percent for engines and couplings.

suppression became nationwide policy, and by the 1940s Smokey Bear was appearing on millions of posters telling the American public, "Only you can prevent forest fires."

Smokey Bear won, but forests around the country—and many endangered species—lost out. Fires were indeed suppressed for half a century or more. Nationwide, creatures that need fire-maintained, early-succession habitats declined in numbers. And without moderate fires to burn them up, fuel loads simply grew greater. When blazes did eventually break out (as in the great Yellowstone fires of 1988) they were correspondingly larger, hotter, and more devastating.

In the longleaf forests, the prolonged absence of fire allowed hardwood trees to grow into a dense mid-story layer, suppressing the growth of pines and turning open park landscapes into closed woodland. For RCWs, already restricted to a forest that had been reduced in area by over 95 percent in just 200 years, this meant that even what was left was becoming less habitable. The new oak mid-story provided rat snakes and other predators with easy routes to their nest holes; the fragmentation of their habitats exposed them to increased conflict with less specialized nest hole thieves like flying squirrels and the much bigger pileated woodpeckers; and the shortage of old trees and consequent increase in distance between cavities made hole defense still harder. By 1973, perhaps less than 0.5 percent of the original million-plus groups of woodpeckers remained, and the RCW became one of the first species to be listed under the Endangered Species Act.

SSS in Action

ESA listing was obviously intended to help the woodpeckers out, but on private land at least, it had the opposite effect. This mattered because, although three-quarters of all remaining RCW groups lived on federal or state land, these populations alone were believed to be too small and too isolated to ensure recovery. Private land was a vital part of the jigsaw. Yet listing meant that for private landowners, RCWs were bad news.

For those wanting to harvest their longleaf, ESA status for the woodpeckers meant an immediate economic burden. Fish and Wildlife Service regulations required that within 70 meters or so of each nest hole, no large trees[12] could be cut, no roads could be built, and no pesticides could

---

12. Where a large tree is defined as one over 10 inches (about 25 centimeters) in diameter, measured at breast height.

be used. Not only that, but owners had to leave a 24-hectare foraging area nearby, and at least 280 square meters of large trees (measured in terms of their cross-sectional area)—the equivalent of thousands of otherwise marketable trees, for every group of birds. The costs to landowners were potentially huge—anywhere between US$30,000 and US$200,000 per woodpecker family.

Ironically, ESA restrictions presented a problem even for those landowners who had previously preferred not to cut down their longleaf but instead to manage it just as the woodpeckers like it. Following RCW listing, they became afraid to continue to do so for fear of having increased restrictions placed on their land. Annually raking up fallen longleaf needles (known as pinestraw) for sale as garden mulch can be as lucrative as selling timber. Yet this requires regular burning to keep the mid-story down and the forest floor accessible—much as the woodpeckers prefer. Likewise, managers of golf courses, of which there are very many in the Sandhills, like to cultivate open, park-like pinewoods. So too do foresters who are managing their plantations for hunting quail, deer, and turkey. But each of these practices risked attracting more woodpeckers and hence more ESA restrictions on future options—so instead landowners chose to limit management of their forests, thereby making them less appealing to RCWs.

Unsurprisingly, resentment of both woodpeckers and Fish and Wildlife Service officials became rife. Dougald McCormick Jr., pipe-smoking wit and owner of a prime tract of longleaf in North Carolina, where his family settled 200 years before, captured the mood when he got a license plate for his pickup that read "I EAT RCWS." The red-cockaded woodpecker had become the northern spotted owl of the South.

Jay Carter, longtime RCW biologist suddenly caught in the crossfire of ESA listing, recalls, "Woodpeckers had the whole of the southeast pretty much paranoid for 10 or 15 years. People were really worried they would make their land useless." The last thing landowners wanted was to let Jay onto their property to look for the birds. One of his colleagues tells me, "If Jay turned up, they'd meet him with him a shotgun." Another remembers, "We were so unpopular we used to have our cars driven off the road." One landowner in particular decided to make a very public stand.

When Ben Cone's grandfather bought 3,000 hectares of deforested land along North Carolina's Black River back in the 1930s, the absence of trees prompted his friends to label it "Cone's Folly." But decades of careful management produced open pine parkland that was perfect for hunting quail—and coincidentally, rather good for RCWs, too. By 1991 the property supported 12 family groups of woodpeckers, so when Ben Cone decided he wanted to harvest some of his now-mature trees, the Fish and Wildlife Service instead told his forester that he had a legal requirement to maintain the trees on around one-eighth of the land. The service offered Cone the option of an "incidental take permit"—a provision of a 1982 amendment of the ESA—that would have allowed him to log the area provided he underwrote the restoration of habitat elsewhere. But he refused, and instead filed a high-profile lawsuit against the government, claiming it had deprived him of US$1.4 million in lost timber revenue. He also took preemptive action against what he called further "infestation" of his land by RCWs, clearfelling around 240 hectares of as-yet unoccupied pine.[13]

Plenty of others also went in for preemptive harvesting of suitable but woodpecker-free habitat. Questionnaire surveys showed that between the mid-eighties and mid-nineties, landowners living near known RCW populations were about 25 percent more likely to harvest their timber, harvest young-aged stands, and clearcut their land than were landowners who lived further from RCWs and so felt less vulnerable to woodpecker colonization. Repeated measurements of over 1,000 privately owned forest plots confirmed the questionnaire responses. Within North Carolina's Sandhills region alone, 6 years of preemptive harvesting removed sufficient vacant but suitable habitat for as many as 70 groups of woodpeckers. For RCWs (and everything else that depended on open longleaf forests), the law of unintended consequences had become a reality.

Between them, preemptive harvesting and unfavorable management to avoid the perceived risks of keeping longleaf stands open greatly worsened the woodpeckers' woes. Despite ESA protection, woodpecker numbers plummeted: in South Carolina by over 50 percent between 1977 and 1989, with the rate of decline on private property roughly twice that elsewhere. By 1990 there were fewer than 3,500 breeding pairs left nationwide, and prospects for recovery looked bleak.

Fashioning a Carrot

"At first we thought landowners needed economic incentives," says Michael Bean, "but they soon told us the real problem was the fear that attracting woodpeckers would place new restrictions on their land." Bean—tall, care-

---

13. Cone eventually settled out of court, paying US$45,000 for his "incidental take permit," which allowed him to destroy the habitat of 12 breeding groups. But in exchange he supported the conservation of 21 RCW groups on properties elsewhere in the state.

fully spoken, and by his own self-deprecating description "a recovering lawyer" but in the opinion of others the country's preeminent authority on wildlife legislation—recalls what he discovered in the early nineties when he first got involved in the woodpecker wars. "Instead of encouraging land-owners to do good things for endangered species, we found the regulations were having the opposite effect. We saw that what was needed was a mechanism so that owners could manage their land better without incurring new impediments."

The organization Michael works for,[14] Environmental Defense Fund, built its reputation on landmark legal cases forcing major organizations (including different arms of government) to comply with the law. They helped bring about the nationwide ban of DDT in 1972, made the case for the Safe Drinking Water Act in 1975, and a decade later convinced legislators of the need to phase out lead additives from gasoline. "Sue the bastards" became the house motto. "Like many environmental NGOs, back then we believed the only tool was a hammer," he explains. "But if that's all you have, suddenly a lot of issues begin to look like nails."

Bean and his colleagues decided to add more tools to the toolkit. In the late 1980s, Environmental Defense Fund helped devise the "cap-and-trade" mechanism of the Clean Air Act. This tackled the acid rain problem by forcing sulphur dioxide polluters to reduce their emissions—but with the groundbreaking idea that they could trade pollution allowances. This meant that older factories that would be very costly to clean up could instead meet their emissions-reduction targets by paying other, newer factories that could be improved far more cheaply to lower their emissions way below their own targets. And what happened when a regulatory stick was combined with a free-market carrot? Overall emissions fell faster than the law required at less than one-third of the anticipated cost.

Michael believes that nature conservation needs similarly flexible, incentive-based approaches. "Conservation has mostly been done through confrontation, in courtrooms, and by putting fences round things. But we need to work with landowners too."

The issue that initially drew Michael's organization to the Carolinas was a 1990 Fish and Wildlife Service ruling that Fort Bragg, an enormous military installation in the heart of the Sandhills, was in jeopardy of failing to meet its ESA obligations. Bragg hosts 1 of 10 "recovery populations" for RCWs nationwide, and under the act it was required to build up a popu-

14. . . . Or did when we talked. He's since been appointed to a senior job in the Obama administration.

lation of 350 family groups. Yet a combination of poor fire management, Army training activities, and inadequate control of mid-story trees meant woodpecker numbers were heading in the opposite direction. The plan eventually agreed for solving the problem involved improving habitat management on the base, bringing more land under government control, and ensuring more movement among woodpecker populations on public land. But as Bean realized, this required healthy numbers of birds on intervening private land—where they were woodpecker *non grata*.

Michael Bean, Jay Carter, and a handful of others formed a small group that began to focus on the private lands dilemma. Jay brought up the problem facing the Pinehurst Country Club—a world-famous golf complex that had recently improved their longleaf management for golfers (and hence RCWs), and as a result seen new groups of woodpeckers set up on their land. In the eyes of the law this not only increased their own liability under the ESA; if any of the birds dispersed next door, that of their neighbors increased, too. Surely, Jay argued, the ESA should in some way be able to protect rather than penalize Pinehurst and others for doing the right thing.

Michael agreed with the argument, but also realized that in principle the protection that Jay was calling for already existed. Amendments to the act allowing for incidental take permits (like that offered to Ben Cone) meant that landowners could "take" an endangered species on their land provided they mitigated its loss somewhere else. If this was permitted, Michael reckoned, then surely landowners should also be allowed to "take" in the future any individuals that through benign management they added to the current, baseline population on their land.

The basic concept of this emerging "Safe Harbor" idea, as group member and local attorney Marsh Smith dubbed it, was compellingly simple: in exchange for undertaking sympathetic habitat management now, landowners would be free, if necessary, to undo at a later date the results of that good behavior without fear of prosecution. They would still be committed to maintaining habitat for the pre-agreement, baseline population, but not for any extra individuals that arrived afterward. Yet even if everyone reversed their good actions, the populations would still be protected at their original levels, so there would be no net loss; and provided at least some landowners continued to tolerate their extra RCWs, the overall result would be a win for woodpeckers. Michael's expertise convinced the Fish and Wildlife Service as well as Interior Secretary Bruce Babbitt that the ESA could be interpreted this way, and in April 1995 Safe Harbor was launched.

Within weeks, the Pinehurst club signed up. "When they asked if we wanted to become the first enrollee, I said we'd love it," head of golf course

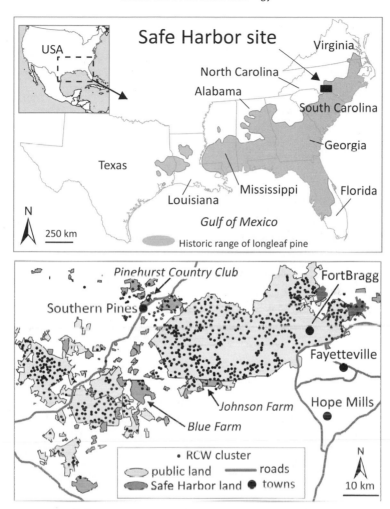

Map 3. In the North Carolina Sandhills, private properties enrolled in Safe Harbor provide vital woodpecker habitat between Fort Bragg and other public land to the west. Map data provided by Susan Miller. Cartography by Ruth Swetnam.

management Brad Kocher tells me as we stand next to an RCW nest tree on the edge of the eighth hole. "At last we could enhance habitat for the woodpeckers and not be penalized for doing the right thing." In exchange for that piece of mind, Pinehurst planted new longleaf trees, kept nest areas clear of hardwood encroachment, and even allowed Jay to drill artificial nest cavities into unoccupied pines.

The result? Inviting fairways edged all around by mature pines. Or if you're a bird, a wide open longleaf woodland with plenty of old trees and virtually no mid-story. Woodpecker heaven, with added golf carts. The woodpeckers responded. The number of family groups rose from the pre–Safe Harbor baseline of 8 to 11, and the birds fared so well that their high-pitched *shrrit* calls even provided the background noise to TV coverage of the 1999 U.S. Open. "Those three extra groups above the baseline," Brad explains, "in the past they'd have been a liability. But they're our Safe Harbor birds, so we're not liable for what happens to them. We've been able to do the right thing without creating a burden for ourselves or our neighbors."

Well-known Sandhills landowner Jerry Holder was another early convert to Safe Harbor. As someone who relied on pinestraw harvesting from the forest floor for his living, he had two worries: keeping his longleaf free from hardwood encroachment without increasing his liabilities from woodpeckers and doing something about pinestraw poachers—nocturnal rustlers who were driving into the woodlands and stealing the needle-growers' crop. Michael was able to help on both counts. "At first we thought pinestraw rustling was a joke, and the local law enforcement officers dismissed it too. But it was a real problem." He and a fellow Environmental Defense Fund lawyer persuaded the sheriffs to take pinestraw rustling seriously—and so won Jerry Holder's trust. "And after he realized we would go to bat for him, Jerry became interested in working with Safe Harbor too."

Not only that, but Jerry went on to encourage fellow landowners to sign up. Because he was a past president of the North Carolina Pine Needle Producers Association, where he led, others followed. The same was true in the golfing world: once such a venerable institution as Pinehurst decided Safe Harbor was a safe bet, other courses followed suit and joined the program. Dozens of landowners and thousands of hectares were enrolled in the first year of the Sandhills pilot. Safe Harbor was up and running.

Harbor in a Storm

Fish and Wildlife Service biologist Susan Miller turns the handle to extend the telescopic pole, and then delicately maneuvers the tiny video camera mounted on its end into the woodpecker cavity 3 meters above our heads. It's early in the season, and Susan's brought her "treetop peeper" along to check woodpecker nest holes for first signs of breeding. An earlier cavity yielded a rather startled flying squirrel (much to my delight, if not the squirrel's); this time the camera displaces a distinctly grumpy screech owl. But no sign yet of woodpeckers.

With us is landowner Julian Johnson. A gently spoken, slim man whose family have farmed here since the 1800s, Julian has a large stars and stripes flying outside his home, a son serving in the military, and an air of quiet authority. He has also become a pivotal figure in Susan's efforts to spread Safe Harbor across the Sandhills. He tells me how he earns his living: "I started out as a rowcrop farmer, with some timber, and got into pinestraw part-time in 1985. Now it's the other way round. The farming is part-time, and pinestraw is my main business."

But the need to keep his longleaf stands open for pinestraw raking brought with it worries about woodpeckers. His family had already lost hundreds of acres to the federal government as part of the compulsory land acquisition program for Fort Bragg, and they were in no mood to lose more. "When the RCWs first came to prominence, we were real concerned that the government would find woodpeckers on our land and take control. A lot of people cut down their trees to discourage the birds. We didn't, but we crossed our fingers and hoped they wouldn't arrive." The change came when he met Susan and her colleague Pete Campbell. "They told me about Safe Harbor. They were practical-minded, and listened to what we needed. In the end I decided you just have to trust someone somewhere. Safe Harbor sounded like a good thing."

Julian joined the scheme in 2000. In exchange for the assurance that any new woodpeckers setting up home on his land won't add to his liabilities under the ESA, he fells fewer of his old-growth pines, keeps the areas between them free from turkey oaks and blackjack, and lets Susan and her team put in artificial cavities to boost RCW numbers. "It requires a bit of compromise on both sides," he explains. "We've changed our straw and timber harvesting practices a bit, and there are still some areas where we want to rake a lot of straw and so don't burn as often as maybe we should. But we manage for wildlife as much as we can. And when you look at the whole together, it looks pretty well."

Importantly, Julian also gets help from a cost-sharing program. "Just recently I got a grant from the Fish and Wildlife Service to brush-cut a 22-year-old longleaf stand. Taking out the mid-story like that will help the woodpecker. But guess what? It'll help with pinestraw production as well."

Improving the pinestraw crop has been the entry point for Safe Harbor on the nearby Blue Farm too, where Susan introduces me to Jim Gray. He's a burly professional forester who manages about 1,400 hectares here and almost 5 times as much across the Sandhills as a whole. "When Susan and Pete arrived here I'd say the Fish and Wildlife people were some way behind the eight ball[15] with the local landowners," Jim says in a strong southern drawl. "They really needed to come up with something besides just a regulatory hammer. We were sitting on a goldmine here in pinestraw income, but what's best for producing pinestraw is also ideal for RCWs too. I knew anything I did to open up the habitat would bring the woodpeckers in."

Again, though, Susan was able to persuade Jim and the farm's owners that Safe Harbor would allow them to boost pinestraw without any added risk from increased numbers of woodpeckers. Walking through a recently burned longleaf tract, with the fresh shoots of resprouting wiregrass vibrant green against the charred floor, she explains that getting the Blue Farm onboard was particularly significant. "It's the largest piece of privately owned land in the Sandhills, and a vital link between Fort Bragg and other public lands to the west. When the Blue Farm signed up to Safe Harbor, they had a baseline of 5 RCW groups. Now that these birds are secure, we're working hard to attract more."

We come to a section that hasn't burned for a few years. On my map it's labeled "Blue E01": an area where Susan and Jim think their first new group should show up. The ground cover is suddenly very different, knee-high with tufts of lank, yellowed wiregrass. The sky behind us is dark and brooding; there's been a tornado 200 miles away in Richmond, and the breeze is strengthening. Ahead there's a longleaf with two holes in the trunk. The top one is a trial excavation by a woodpecker. The lower hole is part of an artificial cavity, installed there last month because of the birds' obvious interest in the neighborhood. The area around is open, park-like, ideal for RCWs. Susan congratulates Jim: "Well, if I were a woodpecker, I'd move in."

And then she spots what she's been looking for. A couple of fresh resin wells close to the entrance to the new cavity. The work of a woodpecker staking its claim. "I think we've got ourselves a new resident. That is really great." She and Jim exchange broad smiles. A discovery that just a few years back would have been a source of conflict is instead a cause for celebration. As Jim wryly remarks on our way back to the road, "We've come a long way since the time when Susan would have been met by a man with a gun."

Later, Susan explains how Julian and Jim, like Jerry Holder and the folk from Pinehurst, have helped spread Safe Harbor. "Take Julian—he's got great standing in this community. Because he thinks the idea's okay, oth-

15. For non-US readers, this translates roughly as being in a tight spot. The phrase has its origins in eight-ball pool (where getting stuck behind ball number 8 puts a player in a difficult position), kelly pool (where being assigned the eight-ball means you have to sink more than the average number of balls), or pool in general (because the eight-ball is black and therefore harder to see than the rest). So that clears things up.

ers have realized it can't be so bad." Julian has gone further, negotiating with The Nature Conservancy, a major NGO, what's called a conservation easement on part of his property, forfeiting his development rights in perpetuity in exchange for a cash payment. This handover of control means that the Johnson land can now count toward Fort Bragg's recovery target of 350 RCW groups. And he goes out of his way to advocate Safe Harbor to other landowners. "My personal campaign has been to prove to others that they can manage their woodland for the benefit of endangered species and still be able to make a living."

The gentle art of persuasion seems to be working, even in unexpected quarters. Robert Bonnie, who works with Michael Bean at Environmental Defense Fund, recalls an early forum for Sandhills landowners and their managers. "We had no idea who would show up. I gave my talk, and then one of the foresters stood up and said, 'You can't trust the Fish and Wildlife Service.' But his employer wanted to know more, and eventually looked over and said 'Aw, hush up!'" The employer's name? Dougald McCormick—the man with the provocative "I EAT RCWS" license plate. Remarkably, he too was persuaded by the logic of Safe Harbor and enrolled his property into the program.

In its first 10 years, no fewer than 91 Sandhills landowners did the same thing, bringing almost 20,000 hectares into Safe Harbor. Some, like Julian Johnson, have signed up for permanent easements of their development rights or even sold land to conservation organizations. Many private properties are now attracting new groups of RCWs above their baseline: great news for woodpeckers without being the liability to landowners they once would have been. And with improved management on their own estate, and with the birds on easements and new NGO land now eligible to be counted, Fort Bragg announced in 2005 that it had met its ESA target of 350 family groups of woodpeckers—a major milestone on the road to nationwide recovery of the species, reached 6 years earlier than originally thought possible.

So can a tool built for woodpeckers help other endangered species? Well, not always. Safe Harbor is obviously not much use for species that don't occur on private land, or for animals (like wolves or grizzlies) that wander so widely that their conservation would require coordinated support across vast areas. But it can work well for relatively sedentary species whose recovery hinges on active help in restoring or maintaining habitats or reintroducing populations—that's where a landowner's support is vital. It turns out that lots of imperiled creatures fit into that category, and so to see how Safe Harbor is helping some of them, I traveled across the country to the great Central Valley of California.

Elderberry Wine and Mark Twain's Frog

Driving through its seemingly endless agri-deserts—one vast, flat, mono-cultural field after another, then another, and another—it's hard to imagine that the Central Valley was once dominated by a series of immense, seasonal, snow-fed wetlands. Erratic rainfall limited farming efforts here mostly to cattle ranching. Then, starting in the 1930s, the largest drainage and irrigation program then seen succeeded in transforming nearly the entire valley into just about the most productive agricultural region on earth. Tomatoes, grapes, almonds, apricots, cotton, avocados, oranges, water-melons . . . one-twelfth of the entire nation's farming output on less than 1 percent of its land.

One of the many bits of collateral damage from this boom was a localized and for the most part invisible insect called the valley elderberry longhorn beetle (less poetically but more practically referred to as the VELB). As their name implies, VELBs live only in the valley and like elderberry bushes. In fact, they're entirely dependent on them—the eggs hatch on the bark, and the larvae then tunnel into the stems, spending up to two years chomping the insides of the elder, before exiting, pupating, and finally emerging as black-and-red adults to wander the outside of the plant for a few weeks in search of a mate.

Such a specialized lifestyle was fine as long as the valley had plenty of native riverside vegetation containing plenty of elders. But the advent of industrialized agriculture led to the loss of over 90 percent of the riverbank woodland, and in 1980 the VELB was formally listed as threatened. With VELBs themselves hard to see or track, their host plants became a target of conservation. But just as in the Sandhills, landowners who still had elders on their properties began to grow jumpy about their liabilities under the Endangered Species Act.

"People wanted to avoid the Fish and Wildlife guys at all costs," recalls articulate young grape farmer Aaron Lange while he shows me around his family's carefully tended vineyard in the heart of the valley. "They'd fire a warning shot over their heads if they came onto the property. And if they found an elderbush, they'd go back at night and take it out." But Aaron's father and uncle, pioneering winery owners Brad and Randy Lange, who had already restored towering valley oaks to their Mokelumne River estate and trialed new ways use less pesticide without lowering harvests, were determined to buck the trend and put some of their riparian vegetation back.

Reaching a large, open section of ground between the regular ranks of vines and the sluggish river, Aaron explains how, ironically, ESA legislation almost prevented the project from going ahead. "We planned to restore 15 acres of woodland, but one of the species we wanted to plant was the elderberry that hosts the VELB. We said to the Fish and Wildlife Service that while we wanted to help, we couldn't be held responsible if some of the bushes died. At first, they said bad luck—that's still our responsibility." But with the Langes' persistence and news of the Sandhills model spreading, attitudes began to shift, and eventually a Safe Harbor for VELBs was drawn up.

There are now over 200 new elders—on top of the farm's baseline of its original 11 plants—scattered among the saplings of Oregon and live oak, cottonwood and boxelder maple on the restoration plot. Biodegradable mesh cages protect the tender youngsters from deer and beavers while the Safe Harbor agreement protects the Lange family and their neighbors from further liabilities if VELB numbers rise. Again, cost-sharing grants have helped—retiring productive vineyards is an expensive proposition—but even more important, Aaron thinks, has been the change in culture at the Fish and Wildlife Service. "Thanks to Michael Bean's scheme, they now recognize farmers are meeting them halfway across the table. They see we're trying to do the right thing, for the right reasons." And the Langes' leadership is paying off. Other vineyards along the Mokelumne are joining the scheme, and with elder numbers increasing, there are proposals to take the VELB off the threatened list.

Up the road in the rolling rangelands of Alameda County, I find out how Safe Harbor is helping another landowner do the right thing. "Hee hee—it worked!" chuckles lifelong rancher Connie Jess as we count tadpoles in the margins of her recently restored stockpond. Swallows and martins skim the surface, while a red-winged blackbird scolds us from a nearby fencepost. Man-made ponds like Connie's have become especially important refuges for amphibians like the California red-legged frog[16] and the California tiger salamander—once abundant but now both ESA-listed due to human modification of their streams and the spread of newly invading predators like bullfrogs and bass. The latest problem is that many hundreds of the stockponds around the Central Valley fringe are now 50 years old and have silted up. The five-figure costs of dredging and reprofiling them and the six different permits that are needed before restoration can start have made the alternative of simply installing troughs and solar-powered pumps tough for hard-pressed ranchers to resist.

16. Celebrated in Mark Twain's *The Celebrated Jumping Frog of Calaveras County and Other Stories* (1867).

But a new Safe Harbor agreement has streamlined the paperwork to a single form, frozen the landowners' ESA liabilities at baseline levels, and freed up farming subsidies to help with the costs. "We got US$14,585 to excavate our pond," say Connie. "In the cattle business, if you're managing to break even you're doing pretty well, so that kind of money is hard to come by." While we talk, tadpoles skitter in the submerged hoof prints where, in a few weeks' time, the salamander larvae will take up residence. Connie shows me more of the pond's delights—where she's planted willows and where the avocets nest. "Next month I'm going back to frog class, to learn to identify all the frogs. I went last year, but I want to go again."

Rewriting the Law of Unintended Consequences

Safe Harbor has spread out right across the range of the red-cockaded woodpecker too. Three hours' drive from the Sandhills and deep into South Carolina, where roadside signs advertise "Loans, Guns and Gold" and proclaim "When All Else Fails, Turn to God!" I meet up with forester Lamar Comalander. He has the strongest southern accent and the most polite manner I've ever come across, and he's brought more RCWs into Safe Harbor than anyone else. As in the Sandhills, as in California, here in South Carolina it's been a case of trusted leaders, and patient persuasion.

Lamar's biggest contract is to advise on the management of Brosnan Forest—an immense plantation owned by the Norfolk Southern Railroad and used for entertaining high-ranking employees and their corporate clients. For over 30 years Lamar has planted new longleaf stands, used burning and brush-cutting to open up century-old ones so they're better for quail and turkey hunting, and along the way created ideal conditions for the largest woodpecker population on private property. He made sure Brosnan was protected by the South Carolina Safe Harbor program as soon as it was set up and has since used cost-sharing grants to help add 12 new groups to its baseline of 67 RCW families.

Lamar has also brought dozens of his other clients and the 100-plus RCW groups on their land into the program, and yet more landowners have followed his well-respected lead: plantations, a chemical plant, even a Trappist monastery. By 2005, 160,000 hectares had been enrolled across the state, with 20 above-baseline groups established as a result. And Lamar has also piloted a bold and, for some people, controversial twist to Safe Harbor: woodpecker banking.

Echoing in some ways the successful cap-and-trade system for sulphur dioxide pollution, the idea behind this sort of "mitigation banking" is to allow landowners with thriving populations of imperiled animals or plants to be rewarded with payments from would-be developers who are then permitted to "take" the same species on their own property. In the context of Safe Harbor, program enrollees become banks by adjusting their baseline upward—in effect taking on full ESA responsibility for some of the extra, above-baseline individuals on their land.

The system—already in place for 40-odd species outside Safe Harbor—works for developers, because land values and hence the costs of conservation vary widely (so, like older factories paying newer ones for disproportionate pollution reduction, it's cheaper for them to pay for conservation elsewhere than to forego developing their own land). It works for the conservation bankers, because they get paid for what many of them were doing anyway. And to the extent that it substitutes poorly protected and isolated populations for same-size (or larger) additions to more robust and better protected ones, it can help endangered species too. Lamar sees banking as a powerful way to attract new landowners into Safe Harbor; he's already brokered a US$100,000 agreement for one South Carolina property to add an extra RCW group to its baseline and is negotiating a second deal for twice that amount. Clearly, good stewardship can be good business too.

Unconventional banking aside, there are now Safe Harbor agreements for red-cockaded woodpeckers across pretty much all their range: in Texas, Georgia, Virginia, Louisiana, Florida and Alabama. These have spawned 70 new family groups of birds, so that after more than a century of decline, many RCW populations are now stable or increasing.

Elsewhere, Safe Harbor has persuaded Bob Long—a self-declared "gun-toting redneck Texas Republican preacher"—to encourage endangered Houston toads onto his property, and many of his fellow Texans to manage their land for aplomado falcons, ocelots, or golden-cheeked warblers. There are Safe Harbors for swallowtail butterflies in Florida, gopher tortoises in Mississippi, prairie dogs in Utah, and even for northern spotted owls in California. All told, the scheme now helps protect 76 species, covers 1.7 million hectares of land, and involves more than 400 landowners—not just farmers, foresters, and ranchers, but water authorities, Native American groups, and even a Boy Scout organization.

There are, of course, limits to what Safe Harbor can achieve. In the case of the RCWs, there are persistent threats (beyond the reach of Safe Harbor) from suburban sprawl, public unease with widespread fires (even though regular burns reduce the risk of uncontrolled fires), and homeowners' well-intentioned fondness for garden bird feeders (which boost numbers of nest hole competitors like flying squirrels). More generally, Safe Harbor itself faces important challenges—in particular, getting greater financial assistance for landowners to help with the costs of management and cutting down on red tape so that bureaucracy isn't a barrier to participation.

But despite these caveats, it is evident that Safe Harbor is already a major success. One reason is it's based on a clever bit of lateral thinking—considering the problem of endangered species on private lands from the perspective of the landlord and then designing a way of overcoming the perverse disincentives created by the ESA. On the ground, its achievements can be credited to patience and to the engagement of leaders in their communities—Julian Johnson and Brad Kocher in the Sandhills, Lamar Comalander in South Carolina, and the Langes in the Mokelumne River winelands. Above all, the success of Safe Harbor has been based on a shrewd bet—on the idea that, given the opportunity, people in general will do the right thing, and they won't change their mind later on.

And when you think about it, the way it's been constructed makes Safe Harbor a fairly safe wager. Under the worst case, if every landowner leaves the scheme, the species is back where it started: no net loss. But even if that eventually happens, in the meantime—as a basic condition of any agreement—baseline populations are being properly managed (rather than neglected), numbers are growing, with individuals dispersing into new areas, and there's a breathing space to search for other solutions. Most importantly, however, it turns out that landowners really don't change their minds. Out of the 400-plus who have so far signed up to Safe Harbor nationwide, only a handful have left it, and most of those have only done so because they've sold their land to conservation organizations. The assurance of no increase in their legal liability, plus maybe some help with the costs, are all that many landowners need to manage their land better for endangered creatures, and even welcome a few more of them onto their property. That's a striking result.

By challenging and changing attitudes—among landowners and, as Aaron Lange made clear, among the staff of conservation agencies as well— Safe Harbor has begun challenging the paradigm of conservation in the United States. As Michael Bean says, "We still need regulatory tools, but we need cooperation too. We're finding that for working landscapes it's often possible to do farming, to do forestry, in ways that share the landscape with biodiversity." This same sort of imaginative pragmatism—finding fresh solutions to old problems by discovering new partners and broader motivations for conservation—underpins most of the rest of the stories in this book. But nowhere more so than in South Africa.

Figure 1. Norman Moore, savior of the Dorset heathland. Photo courtesy of Norman Moore.

Figure 2. The successful reintroduction of wolves to Yellowstone has had profound knock-on effects through the ecosystem. Photo by Tom Collopy, Wild North Photography.

Figure 3. In Micronesia, innovative conservation NGO Rare's radio soap opera *Changing Tides* brings social and environmental issues to widespread attention. A quarter of listeners say the show has prompted them to stop eating sea turtle eggs. Photo by Paul Butler, Rare.

Figure 4. This snow leopard photographed itself by inspecting a camera trap above the Spiti Valley in northern India. Photo by Charudutt Mishra, Nature Conservation Foundation and Snow Leopard Trust.

Figure 5. Poaching in the raw: this female took four days to die after her horn was hacked off by poachers. Her calf was also killed. Photo by Uttam Saikia, Bhumi.

Figure 6. Rhino poacher Golap Patgiri with his son Bharat. Photo by Andrew Balmford.

Figure 7. Female red-cockaded woodpecker. Photo by Eric Spadgenske, U.S. Fish and Wildlife Service.

Figure 8. Richard Cowling: botanist, surfer, and one of the main thinkers behind Working for Water. Photo by Shirley Pierce.

Figure 9. Fynbos is richer botanically than almost any other habitat on earth. Photo by Andrew Balmford.

Figure 10. Hacking down unwanted Australians. Photo by Andrew Balmford.

Figure 11. Koniks thundering across the Oostvaardersplassen. Photo by Hans Kampf, http://www.grazingnetworks.nl.

Figure 12. The immense Crailo ecoduct nearing completion. For scale, see the train on the right-hand side. Photo by Willy Metz, Goois Natuurreservaat.

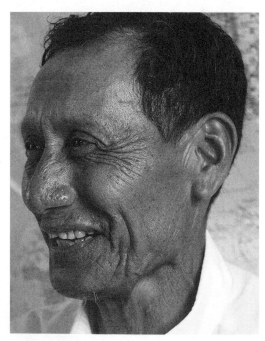

Figure 13. Don Alejandro Ramírez, one of the Loma Alta elders responsible for the *comuna*'s decision to set aside much of its land as an ecological reserve. Photo by Andrew Balmford.

PROBLEM PLANTS,
POLITICS, AND POVERTY

Something astonishing is happening in South Africa, which shows how having the wit to see problems through others' eyes can open up conservation opportunities on an immense scale. Politically savvy ecologists deeply concerned about the threat that exotic plants pose to the country's exceptional floral diversity have persuaded a cash-strapped post-apartheid government to invest hundreds of millions of dollars in their removal—not out of concern for biodiversity but because the aliens also drink a lot of scarce water, and their laborious elimination provides jobs and training for thousands of unemployed people.

Walking away from the remote mountain track to admire the spectacular 400-meter-deep gorge opening up to our left, my guide and good friend Richard Cowling bends down and unceremoniously uproots a Monterey pine seedling, snapping it in half. It's a curious scene. With his shoulder-length hair and year-round tan, Richard looks more like a surfer dude than an internationally renowned conservation scientist (he is in fact both). And *Pinus radiata*, to give the plant its scientific moniker, is restricted to just a handful of sites in its native California.

But we're not in Monterey County. We're in the rugged Bosberg Mountains of South Africa's Eastern Cape Province, where an unbroken blue sky lights impressive peaks and Monterey pines keep conservationists like Richard awake at night for a very different reason: they're among the worst of around 200 nonnative plant species now invading South Africa, drinking vast quantities of much-needed water, damaging property and farmland by increasing the severity of wildfires, and posing a seemingly insurmountable threat to the country's exceptional diversity of plants.

South Africa is botanically extraordinary. With over 20,000 species, it has more plant variety than the United States (which is nearly 8 times its size) and roughly twice as much as the whole of Europe. Two-thirds of its plants are endemic—they occur nowhere else on earth. The country has

yielded many of the world's most treasured garden plants—from gladi-
oli, geraniums, and gazanias to arum and amaryllis lilies, bird-of-paradise
plants, and red-hot pokers. Carl Linnaeus, the eighteenth-century Swede
who invented the system of taxonomy still used to classify species,[1] so ad-
mired the region that he described it as "that Paradise on Earth . . . which
the beneficent Creator has enriched with his choicest wonders."[2]

But today Linnaeus's botanical paradise is in trouble. More than 1
in 8 plants native to South Africa are threatened with extinction. As in
other countries, conversion of their habitat for human use is the greatest
threat, but in South Africa exotic invaders are not far behind. By growing,
reproducing, and spreading faster than native plants, they are a serious
problem for over half of the species on the national Red List of threatened
species.

Alien plants are harming South African animals too. The spread of exotic trees into previously open country is favoring some tree-nesting bird
species, but at the expense of many others. By expanding along watercourses, nonnative trees and shrubs are shading out previously sunny
breeding sites of South Africa's dragonflies and damselflies: several species seem to have disappeared completely. And in Africa's oldest nature
reserve—the St. Lucia wetland complex north of Durban—invasive plants
are even threatening one of Africa's oldest and most impressive predators.
In common with a handful of other animals the sex of crocodiles is determined by the temperature they experience as eggs in the nest. Hotter nests
produce more females, but now, increased shading of St. Lucia's riverbanks
by the appropriately named triffid weed is making crocodile nesting sites
5–6°C cooler. There are real fears that as a result, the population will produce too few females to be self-sustaining.

Ecologists reckon that triffid weeds, Monterey pines, and dozens of
other invasive plants already extend over one-twelfth of South Africa,
and that the area they occupy is increasing at around 5 percent each
year. Their spread is worst of all in the African counterpart to the Dorset
heathlands—a spectacularly rich, fire-prone vegetation type called fynbos

1. Linnaeus's own name was itself the product of a new labeling scheme. His dad was an early
adopter of permanent last names (rather than the prevailing system of patronyms, such as
Carlsson or Andersson, based on fathers' first names). He chose the surname Linnaeus after a
giant linden (or lime) tree that grew on his property. Carl went on to invent binomial nomen-
clature, whereby each species has a two-part name comprising its genus and its species: hence
Homo sapiens for our particular species of the genus *Homo*, *Pinus radiata* for Monterey pines, and
Tilia cordata for small-leaved lime trees.

2. From 1772 letter to Ryk Tulbagh, the governor of the Cape, quoted in Cherry, "South Africa—
Serious about Biodiversity Science."

("fine bush").[3] With its clumps of low olive- and sage-colored shrubs, punc-tuated by flowers in yellow or pink, orange or white, fynbos looks at first glance like heathlands elsewhere. But in terms of plant diversity, it's in a league of its own. Deposit a botanist in it, tell them to learn a new plant ev-ery day, and they'll still be going 25 years later. Nearly 6,000 of the species on their career-long list will be found nowhere else. Yet a fifth of the fynbos has already been invaded by alien plants, and mathematical models predict that, if left unchecked, these could overrun more than three-quarters of it by the end of this century.

Faced with a threat on this scale, Richard Cowling's instinctive weed-ing of an exotic pine seems poignant but futile. But the new South Africa is extraordinary politically as well as biologically. In the face of a sobering array of competing social, economic, and health demands on its resources, the country's post-apartheid government has set up and largely funded the biggest alien species removal program in the world, with current in-vestment running close to a staggering US$100 million each year. Why? Because imaginative conservationists calculated not just the biological but the immense human costs of invasive plants and managed to get the infor-mation to decision makers at just the right time. Quantitative conservation science meets realpolitik.

Releasing the Genie

The first European settlers arrived at the Cape in 1652 to find a semi-arid country dominated by the low shrublands of the fynbos, and where a combination of climate and natural fires meant there were very few for-ests. Charged with providing crops and meat to refuel the Dutch East India Company's ships traveling from Europe, the colonists set about introduc-ing crop species from around the world. By the nineteenth century, set-tlers had added North American and European pines and Australian gums and wattle trees. The exotic trees were planted not just to provide timber,

---

3. Like the flora of the longleaf forests, fynbos plants have evolved a suite of adaptations for liv-ing with fire. My favorite is a phenomenon (seen elsewhere too) called myrmecochory, where plants attach nutrient-rich bodies known as elaiosomes to their seeds. The elaiosomes produce chemicals that mimic those used for communication by ants, so when the seeds fall the ants take them to their underground nests, where they consume the elaiosomes. But the rest of the seeds are too hard to eat, so they remain underground, safely out of reach of fires, until conditions are right for them to germinate. Almost a third of fynbos plants go in for myrmecochory, but now another unwanted alien—the Argentine ant, which eats elaiosomes without burying the seeds, and which aggressively displaces the native ants—is disrupting this intricate and vital process.

firewood, and tannin for the leather industry but also to increase rainfall. Sadly, though, trees don't make it rain, and while in some circumstances (as I'll find out later in this book) they can help clean up and even augment flows of fresh water, in areas like the Cape they can literally drink the landscape dry.

Government incentive schemes encouraged tree planting by white farmers across what are now the Western and Eastern Cape provinces and into KwaZulu-Natal. In the vast arid hinterlands of the Karoo, ranch owners introduced prickly pears and mesquite bushes from the Americas to provide animal fodder. Lantana, bugweed, and lilac were brought in from South America and Asia as ornamental plants to decorate homesteads. As late as the mid-1970s, Australian wattles like rooikrans and Port Jackson willow were being planted in enormous numbers in coastal areas to stabilize mobile dunes, while belts of huge Mexican sisal plants were established along the country's borders as spiny-leaved barricades against the perceived threat of communists invading from Angola and Mozambique in Soviet-backed tanks.

The impacts of these introductions have been far-reaching. Today, despite limited rainfall, croplands and forest plantations together cover one-tenth of South Africa and contribute about the same fraction of its jobs. So far, so good. But nonnative organisms—or at least a troublesome minority of them—also have an unfortunate habit of escaping from where they are useful and instead becoming profoundly problematic. South Africa's real invaders turned out to be not red, but green.

Out of over 9,000 plant species introduced to the country, around 200—mostly trees and shrubs—not only have spread but have begun to do serious damage. Wind-dispersed pines have sent seeds out from plantations and across mountainsides. Fast-growing gum trees and black wattles have dispersed down streams and rivers. Prickly pears have invaded large areas of arid grazing land. And rooikrans and Port Jackson willows have become dominant across large tracts of coastal vegetation. So how have these aliens managed to become such a problem?

Most of the worst offenders in this real-world great escape owe their success to a number of particularly effective traits. They set seed and then germinate quickly on ground that's been disturbed by people, livestock, or machinery. Freed from the pests and predators that held them in check back in their homelands, the colonists grow and mature quickly and reproduce profusely. Many also have a clever trick for exploiting the region's naturally occurring fires to gain further advantage over the competition: they like especially hot fires and are even able to promote them.

Whereas most of South Africa's native plants become dormant during the dry season (in order to conserve water), the invaders typically use deep roots to tap ground water and so grow year-round. With no seasonal shutdown, the plants grow bigger. From the perspective of fires, more mass of plants equals more fuel—so the spread of alien weeds leads in due course to much hotter burns. These kill off plants and seeds of native species—even those that are fire adapted—but actively stimulate fresh growth of adult plants and germination of seeds among the invaders. As a consequence, each fire brings yet greater domination by exotics. And as well as fundamentally altering fire regimes in their favor, the invaders also use far more water than the plants they replace.

Aliens Drank My Farm

Concerns about water consumption by alien trees first surfaced in the 1920s, when farmers blamed them for many of South Africa's rivers drying up and petitioned the authorities to investigate. In response, the government set up a research station at Jonkershoek, some 50 kilometers east of Cape Town, where beginning in the 1930s, a pioneering hydrologist named C. L. Wicht built a series of concrete weirs where he could measure the flow rates of different streams. Wicht was especially interested in the effects of plantation forestry on water availability, and so he planted some of his catchments—the areas that drained into his streams—with exotic pines, while leaving others under fynbos.

It took a while for the new trees to grow, but as the results built up they began to provide strong evidence in support of the farmers' suspicions. Instead of leaving water to run downstream, the trees soaked it up and passed it on—via their leaves—into the atmosphere. Streamflows were lower in catchments with pines, pointing the finger of blame not just at the plantations but also at the forestry trees that had spread outside them. By the mid 1970s, Wicht's successor had enough information to calculate that pine plantations and their escapees reduce runoff to streams by more than half, prompting him to warn government that the invasion of the Cape mountains was seriously reducing regional water supplies.

The government had already begun to act. It introduced legislation requiring the land around plantations to be cleared of escaped aliens and created a new Department of Forestry, which meant that for the first time managers, planners, and researchers concerned with water supplies could work alongside each other within the same organization. Large-scale, labor-intensive alien clearing began in earnest, with field teams cutting down exotic trees and shrubs across big tracts of the country. From 1970 to 1974 the area treated doubled, to over 350 square kilometers in a year.

Research efforts intensified too. Experiments in the northeast of the country, in what is now KwaZulu-Natal, revealed that streamflow there dropped by more than four-fifths after catchments covered by open grasslands were planted with pines. Streams in Mpumalanga, a province east of Johannesburg, dried up completely after grassland was replaced with a mixture of pines and gum trees. Both sets of results were from dense plantations close to watercourses, but they probably provided a reasonable indication of what was happening in the alien-choked catchments responsible for providing most of South Africa's water.

As well as making plain the reality of the problem, researchers developed new solutions for addressing it. Invasive plants can prove as resilient to attack as Hydra, the monster with multiple, resprouting heads that the classical alien-slayer Hercules only eventually killed after decapitating, cauterizing, bludgeoning and finally burying it under a large boulder.[4] Adopting much the same approach, scientists in the Department of Forestry developed what they called integrated control programs, which amounted to killing the invaders in as many ways as possible. First, workers felled the plants. To limit regrowth, they also poisoned the stumps. A few months later, a follow-up treatment of careful burning simultaneously removed fuel and killed exotic seedlings that had germinated in the cleared areas. Finally, the work teams returned every year or so to uproot any offspring that had somehow escaped previous onslaughts. Hydra slaying, Africa-style.

Research also confirmed the benefits of all this Herculean clearing. As expected, scientists found that removing exotic trees increased the amount of water left to run on into streams. In one example, within just two days of felling a dense stand of pines and wattles on each side of a stream in Mpumalanga, streamflow more than doubled. In another, removing Monterey pines from one-tenth of a stream's catchment in the Western Cape and allowing the fynbos to regrow increased flow rates in the first year by almost half.

But while technical know-how and scientific evidence for aggressive alien eradication were mounting, political will and organizational mo-

---

4. Coincidentally, Linnaeus had trouble with a hydra too. On a visit to Hamburg he was proudly shown a stuffed specimen of the mythical beast by the mayor, who hoped to sell it. Linnaeus was not convinced and publicly denounced the specimen as a fake, upsetting the mayor's plans and prompting a swift exit from the city.

mentum were being lost. In the mid-1980s political deals struck in the dying days of the apartheid government led to responsibility for catchment management being passed from the national level down to the provincial conservation bodies, who were given neither the resources nor the training to carry out their new tasks. Research was also relocated to an organization called the Council for Scientific and Industrial Research (CSIR), and the close linkage between science and action was severed. The net effect was a steep decline in on-the-ground clearance and the renewed expansion of invasive plants.

Alarmed by this turn of events, a handful of energetic fynbos ecologists—among them Richard Cowling and William Bond from the University of Cape Town, and Brian van Wilgen and David Le Maitre from CSIR—started flagging the costs of neglect. As biologists, they echoed Wicht's warning from almost half a century earlier, that the insidious spread of nonnative plants was "one of the greatest, if not the greatest, threats to Cape vegetation."[5] They suggested that local extinction of many plants was a real possibility. But dire though the situation was, the scientists knew that realistically, biological arguments alone wouldn't be enough. However, they knew something else too: that the invaders were a huge problem for South Africa's people as well as its plants.

So the ecologists reexamined the water numbers. Using Wicht's long-term data they estimated that if the current situation continued, water yield in the Western Cape would drop by nearly a tenth. Further deterioration of funding for alien control might increase this figure to a half. The scientists also pointed out that nonnative species were affecting people via fire, as well. Dense stands of exotic trees were not only fueling more and hotter burns but hampering the construction and maintenance of fire breaks around settlements and restricting access for firefighting. Last, invasions were leading to widespread soil erosion. Whereas intact fynbos vegetation binds the soil, intense fires in heavily infested areas seriously damage it, creating a surface that repels rather than absorbs rainfall. With invasions left unchecked, rainstorms after fires were resulting in rapid runoff and serious soil loss. The associated flood damage to homes, drains, and other infrastructure was beginning to mount up.

In the political upheavals of the late 1980s and early 1990s, none of this

5. From C. L. Wicht, "Preservation of the Vegetation of the South-Western Cape: Special Publication of the Royal Society of South Africa, Cape Town, South Africa," quoted in Van Wilgen et al., "The Sustainable Development of Water Resources: History, Financial Costs, and Benefits of Alien Plant Control Programmes," p. 404.

was by itself enough to change government policy. But concern about water was growing. With its important agricultural base and rapidly growing urban populations, water availability was increasingly being seen as the greatest physical constraint on South Africa's economic growth. The collected warnings of the scientists rang alarm bells at the Department of Environment Affairs and led to funding for CSIR's Van Wilgen and Le Maitre to get a better handle on the consequences of plant invasions for water yields. And that, in due course, catalyzed a dramatic turnaround.

The Power of a Timely Idea

The Kogelberg Mountains are among the most distant peaks in Cape Town's breathtaking skyline. They are also relatively free of alien plants and instead support some of the richest fynbos vegetation in the world. With few invading black wattles, gum trees, or pines to contend with, the slopes of the Kogelberg host more native plant species than occur in the whole of the British Isles (an area 1,750 times larger). No fewer than 150 plant species are found nowhere else on earth, making the area's mountainsides and ravines a magnet for adventurous botanists in search of such rarities as the Kogelberg sceptre, the marsh rose, and the matchstick pagoda. High-quality maps of the extent of fires, different vegetation types, and the progress of plant invasions also made the Kogelberg an excellent place for the CSIR team to develop a state-of-the-art assessment of how the spread of exotic trees might change Cape Town's water supply.

The ecologists began by feeding plant distribution maps and growth data into computer models of how nonnative species might proliferate and disperse after fires. The result was a series of maps of the likely spread of the invaders if left unchecked. These confirmed earlier fears. Even though only 2 percent of the Kogelberg was currently infested, this figure was projected to rise to well over half within a century. Next, the scientists combined these projections with what they knew (from the experiments at Jonkershoek and elsewhere) about the drinking habits of nonnative plants, to estimate how streamflows would be expected to change with the impending invasion. Their calculations told them that, averaged over the next century, annual water flow from the Kogelberg was set to decrease by 35,000 cubic meters per square kilometer. Thats works out at about 2,500 Olympic-sized swimming pools lost each year in the Kogelberg alone. If these results were typical of the nearby catchments that supplied Cape Town, the loss in an average year would be equivalent to one-third of the city's water use. But the relative loss would increase over time, and in drought years and more heavily invaded areas it looked set to be greater still.

A second study then took the results of these models and examined their economic implications. The team looked at how the costs and benefits of building a hypothetical new dam for supplying water would differ, depending on whether its catchment was left to be taken over by nonnative species or if it was cleared and then kept free from reinvasion. Was chopping down aliens worth it? Their figures showed that although clearing increased the annual outlay, this would be more than offset by the gain in water yields: overall, removing exotic plants would reduce the cost of delivering a liter of water by 14 percent—and would yield many millions more liters of water into the bargain. Keeping catchments clean made sound economic sense.

The ecologists now had the evidence they needed. Removing nonnative plants was not only essential for the persistence of South Africa's exceptional biodiversity, it was also the most prudent way of meeting the country's increasing demand for water—and at the same time could provide employment for thousands of semi-skilled workers. This realization of the alignment of conservation with economic and social interests could not have been more timely. In 1994 South Africa had just elected its first democratic government. Politicians were concerned as never before with the problems of how to provide for the 12 million South Africans who lacked access to drinking water, the 20 million without sanitation, and the 8 million without jobs. There was also an unprecedented willingness to try new ideas.

The scientists decided to go straight to the top and target the new and charismatic Minister for Water Affairs, Kader Asmal, a man who until then had reportedly considered his cabinet portfolio to be disappointingly low-key. The ecologists' pitch—emphasizing not just the economic rationale for clearing away alien plants but also its potential to offer tens of thousands of jobs—was spot-on. Showing a degree of spontaneity and imagination that would be remarkable in a less nimble or more cynical democracy, the minister was convinced. Within three weeks of the June 1995 presentation, cabinet released over US$7 million, and Working for Water was born.

Scaling Up

Exciting and revolutionary innovations raise expectations and need to deliver tangible results as swiftly as possible. Working for Water was no exception. Fortunately, from its inception the program was driven forward by an exceptionally energetic management team. Galvanized by its leader, Guy Preston—a man who lives and breathes WfW and seems not to need sleep—the program metamorphosed in just a few months from a cherished idea into an organization running projects across South Africa.

WfW focused from the outset on the poorest of the poor. It hired unskilled and unemployed people—more than half of them women—from impoverished rural communities and trained them to cut alien trees, spray stumps, and pull out seedlings. In its first 8 months the program cleared over 300 square kilometers of invaded habitat (equivalent to 40,000 football pitches), created more than 6,000 jobs, and (crucially in government circles) spent all of its budget. Just as important, it made itself well known. All employees wore dazzling yellow T-shirts, and a dedicated communications project made sure that the government in general—and supportive ministers in particular—got equally visible credit for Working for Water's successes among a public eager for concrete achievements from their new democracy.

Momentum gained on the ground was compounded by the continued accumulation of scientific evidence of the problems posed by alien plants. This was important. As Brian van Wilgen (the pragmatic scientist who headed the CSIR team) explained to me, "convincing a receptive minister was one thing; persuading officials in the Department of Water Affairs and Forestry was quite another. Most of them were engineers who had difficulties coming to grips with the idea that clearing trees could be more cost-effective than building dams. We needed more data."

So Van Wilgen and his colleagues extended their earlier hypothetical analysis of the costs and benefits of clearance to two real-world catchments: one where a new dam was being planned and one that was already delivering water to Cape Town. In the case of the new dam, their results suggested that, as before, combining dam construction with alien clearance would lower costs and increase water yields. But they also showed that simply removing nonnatives from the catchment of the existing dam would yield additional water at only one-seventh of the price of water from a new dam. Cost savings would be steadily eroded, however, if clearing was delayed.

The need for urgent action was further reinforced by the first attempt to gauge the nationwide extent of the invasive species problem. Working with local experts to produce maps of the spread of each of the main species, CSIR scientists estimated that exotic trees and shrubs had already invaded around a 100,000 square kilometers (or one-twelfth) of the country. This

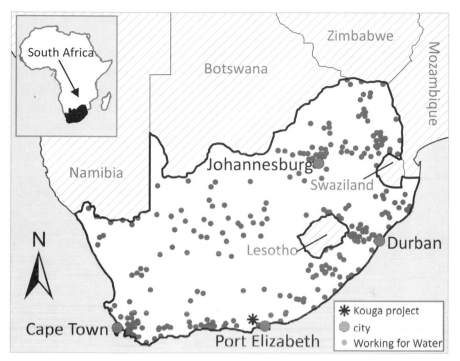

Map 4. Working for Water employs over 25,000 people in around 300 alien-clearance projects scattered across South Africa. Map data provided by Andrew Wannenburgh. Cartography by Ruth Swetnam.

overall figure masked a lot of variation in the density of the invasion, so the scientists also calculated its "condensed area"—how much land the same number of alien plants would occupy if they were concentrated at maximum density. It turned out that South Africa's aliens occupied a condensed area of 17,000 square kilometers—equivalent to a wall-to-wall forest of pines, wattles, and prickly pears covering almost all of Wales or New Jersey. Combining this shocking statistic with the earlier Jonkershoek streamflow results suggested that nonnative plants were now using an astounding 3 cubic kilometers of water—or some 7 percent of the entire runoff of South Africa—every single year.

The government was persuaded. Given the increasingly solid scientific case and WfW's impressive record for spending money efficiently and communicating its results, the program was rewarded with a rapid increase in funding. As well as winning the largest single slice of the government's new fund for poverty relief, Working for Water received levies from water users and large grants from local governments, businesses, and foreign aid donors such as the World Bank. The program's annual budget rose to US$19 million in 1996–1997 and a heady US$54 million a year later.

Since then, there has been no slackening of the pace. By 2010–2011, recurrent funding for WfW amounted to nearly US$100 million per year, with the program annually clearing some 8,000 square kilometers of invaded habitat (one-fifth for the first time, the rest as follow-up). Around 300 projects countrywide employ over 25,000 people. In one sense, Working for Water has become probably the largest long-term conservation initiative in the developing world. Viewed through a different lens, it has become the flagship of the South African government's war on poverty.

Operations on the Ground . . . and from the Air

It's a cloudless April morning, and I'm about to see Working for Water in action. At first meeting, Edwill Moore appears an unlikely champion for the cause. He is quiet and unassuming and has no formal training in hydrology, forestry, or ecology. Edwill was originally hired by the program a decade ago because his army experience meant he could drive the 12-tonne trucks needed for moving workers and equipment. Since then, he's become area manager of one of the 300 Working for Water projects and spends more of his time in a helicopter than a truck, directing the high altitude clearance of the Eastern Cape's Kouga catchment. Edwill has offered to take me up with Roland Black, the Kouga project's regular pilot, for a bird's-eye view of progress to date. Once in the air, the pair's gentle but evident pride in what their crews have achieved and their enthusiasm for the daunting challenges still ahead become infectious.

The scale and enormous difficulties of controlling the alien takeover of the Kouga are quickly apparent. The area is scenically stunning, but vast, mountainous, and in large parts inaccessible by road. Inaccessible to us, but not to the genies we have unleashed on this landscape. The chief enemies here—black wattle, Monterey and cluster pines, Spanish reed, and rooikrans—are obvious even from 5,000 feet. By dispersing their seeds on the wind and in the water they've spread far beyond the plantations, woodlots, and farms where they were originally introduced. Plants from half a world away now blanket the Kouga's once-treeless hillsides and choke its once-free-flowing rivers.

The human costs have been considerable. Water supplies to the fruit-growing farms in the valley bottoms fell. Levels at the dams that store most of the drinking water for the 1.5 million people living in and around the nearby city of Port Elizabeth[6] dropped too. And as the alien plants invaded wetlands, they dried them out and reduced their capacity to absorb excess water—so that ironically, when it rained, flooding became more frequent, sediment loads rose, and water quality declined.

But WfW is now beginning to generate results for the people who depend on the Kouga. The task still facing them is enormous, but from the air Edwill and Roland are able to show me vast swaths—ironically reminiscent of environmental horror stories from Indonesia or British Columbia— where foreign trees lie scattered like spilt matchsticks. On some slopes fynbos plants are starting to sprout again—recovering ground long since lost. And in low-lying spots, verdant native wetlands are beginning to regain the valley floors.

These achievements are down to a workforce of about 600 people, most of whom have never had paid jobs before. Under the rules of poverty-relief funding, each of them is eligible for up to 24 months of employment. Salaries are low—around US$6 for an extremely hard day's work of hacking and sawing and spraying. Low—but just enough to make a real difference to many. People like Katrina Vereen, a colored[7] mother of four from a nearby township who is now a Kouga project contractor. Before Working for Water Katrina had no previous experience with paid employment and little prospect of getting any. But as she proudly explains to me, earning a steady wage by joining and then running an alien-clearing team has enabled her to do something simple yet simply unimaginable before: to send all her children to school.

With such a vast invasive species problem to tackle, setting priorities is all important. In the Kouga, these are driven largely by what will make most difference to water flows. Clearing heavily infested areas near the coast generally receives low priority, as there are few people living downstream to benefit, and because clearance would only be temporary if stands of exotics left further upstream continued seeding the areas. Instead, efforts are focused initially on high-altitude infestations, often reachable only by helicopter, where crews camp out for up to a week. Only when the upland sites are alien free do the teams move progressively down the hillsides. But while this strategy makes sense, the limited public visibility of ac-

6. . . . An area now renamed Nelson Mandela Metropolitan Municipality, after the Eastern Cape's most famous son.

7. In the obsessively regulated world of apartheid, people of mixed race who didn't have quite enough sub-Saharan ancestry to be classified as black were instead termed "colored." Even after the arrival of democracy many people continue to identify themselves as colored.

tivities that are necessarily focused for the time being in upland areas does tend to undermine public support for the Kouga project.

Edwill and his teams face other frustrations too. Stifling health and safety legislation has created legal barriers to the Kouga project using burning as a follow-up tool, even though carefully managed fires would be the most cost-effective treatment for many invasive species. Edwill tries to encourage landowners to carry out burns themselves, but where this is not possible, work crews have to revisit sites repeatedly to poison seedlings and resprouting adults. Some farmers won't accept responsibility for maintaining areas that Working for Water has cleared. A handful of owners have refused to allow the project access to their land altogether: from the sky their uncleared properties stand out like inkblots, dark with still-standing aliens that threaten to reseed the invasion and undo all the hard work. Although new regulations mean that these sorts of noncompliance are technically illegal, much like Safe Harbor's advocates the project managers here adopt a patient approach, working with neighboring landowners to win over reluctant farmers through peer pressure rather than prosecution.

The encouraging news is that many private individuals are increasingly taking on responsibility for tackling invasives. All over the country concerned citizens have formed so-called hack groups to do battle with the alien weeds on their patch. At Cape St. Francis, to the south of the Kouga, enthusiastic planting of Australian rooikrans persisted until the 1970s in a misguided attempt to stabilize naturally mobile sand dunes that rather inconveniently drifted across people's gardens. The rooikrans did indeed stop the sand from moving across the headland, but as a result it was no longer able to replenish the ocean-buffeted beach on the other side of the settlement at St. Francis Bay. As a result the beach is eroding rapidly, and middle-class residents intent on restoring a vital natural process now spend their Saturday mornings energetically uprooting the antipodeans and nurturing the natives.

Fifty kilometers to the west, in another site whose coastal location makes it a low priority for the Kouga project, commercial dairy farmer Ross Naude dedicated two of his farmworkers to full-time clearance of the wattles and cacti invading his valley. He has plans to route a long-distance hiking trail through an exquisite gorge—once blocked by alien plants but where natural swimming pools now brim with crystal-clear water. But he has a more fundamental motivation too, one borne of a profound sense of custodianship for the land. As he puts it, "while the problem is daunting now, it is still just about manageable. If I left it for the next generation, it would simply be impossible."

Spreading the Benefits

Efforts in and around the Kouga are typical of how Working for Water operates on the ground. But the program as a whole is increasingly involving itself in far more than just the mechanical removal of alien plants. At a local level, WfW projects try to offer childcare facilities and set up health and community centers. Employees are taught not just about how to work safely with chainsaws and chemicals but how to handle their finances, manipulate a spreadsheet, and plan for the family size they want. Two key objectives have been to raise awareness about HIV/AIDS and equip workers with the skills they need so that on leaving the program after two years they can compete for jobs elsewhere.

At the national scale, WfW has also been a big supporter of the so-called biological control of invasive species. This involves finding and deliberately introducing their natural enemies—the herbivores and diseases that held them in check back home in their native ranges. These agents are usually insects or fungi, and they do their damage by eating or infecting their hosts, thereby slowing down their growth or reproduction. Almost 100 different biological control agents have been released in South Africa. Their impacts have varied—from outright failure to complete control of the spread of their host species. But even taking the failures into account, biological control efforts turn out on average to be extremely good value for money: each dollar invested in identifying and releasing natural enemies of long-leaved wattles invading from Australia, for example, has resulted in roughly a hundred dollars' worth of increased streamflow.

Besides working on clearance and biological control, Working for Water has been busy promoting secondary industries that can make profitable use of the byproducts of alien removal. The idea is that if these new enterprises take off, then the private sector can create permanent jobs that trained workers can move into once their two years of program funding are over. Activities include selling firewood; making furniture, handicrafts, and environmentally friendly building materials for low-cost housing; and producing charcoal and compost from felled trees. In a development that eloquently underlines the reality of the social problems facing South Africa, WfW has even received a grant from the World Bank to manufacture low-cost coffins from nonnative timber. Under the onslaught of the HIV/ AIDS epidemic, death has become big business in South Africa. Coffins alone can cost families already in financial crisis over US$300. Eco-coffins built from felled aliens sell for one-sixth of that price.

Working for Water has also spawned several other programs that have adopted its approach of explicitly linking better care of the environment with economic and social development. Lethal fires—often made worse by encroaching vegetation—are a very real hazard for the inhabitants of South Africa's townships. In response, Cape Town's Ukuvuka campaign and its nationwide follow-up, Working on Fire, have hired previously unskilled workers to clear wildfire-fueling aliens, create firebreaks, and reduce fire risks in shanty towns. Worries about how invading plants are drying out swamps and marshes have triggered the establishment of another sister program called Working for Wetlands. This has set about building strategically placed weirs to restore wetlands so that they can once again provide free water filtration and flood protection to the people that live downstream of them.

And Working for Woodlands has started looking for novel ways of financing the labor-intensive restoration of other indigenous vegetation. One particularly exciting prospect (picked up again in the appendix) is the possibility of rehabilitating the dense thicket vegetation that covers much of southeastern South Africa. Thicket is botanically almost as precious as fynbos but has been seriously damaged by decades of overstocking with goats. It now seems that long-term funding for its restoration could come from the emerging market in carbon credits. As I discover later from other carbon-saving projects in Costa Rica and Australia, growing numbers of organizations and individuals who burn greenhouse gas–emitting fuels in their cars and businesses or by going on holiday are paying others to recapture the resulting carbon dioxide from the atmosphere. It turns out that regenerating thicket vegetation is extraordinarily good at this. It grows so densely that it can absorb carbon dioxide as quickly as forest growing in places with three or four times as much rain. Depending on how much industrial and domestic polluters eventually pay others to mop up carbon on their behalf, restoring native thicket could end up being more profitable than farming goats.

Taking Stock

These spinoffs give some idea of future potential, but what has Working for Water already achieved? In terms of helping people, since its inception the program has provided more than 17 million days of work. More than half of the jobs have gone to women, one-fifth to young people, and 1 percent to the disabled. Summary statistics can be misleading, but in the case of WfW you don't have to look far to find out how these opportunities have changed real lives. Nomfuneka Booi, a single mother, said, "Before I got this job I was worried because I had no money to buy food and clothes for my daughter. Now I have money, so I don't need to worry."[8] Caroline Gelderblom, a scientist involved in the early days of Working for Water, sums up the program's social impacts with an image from the rural highlands of KwaZulu-Natal: "Before the program started, the children here used to have to walk to school barefoot, even on frosty mornings. Working for Water enabled parents to put shoes on their children's feet."

How far has Working for Water provided water as well as work? Around 20,000 square kilometers of invaded habitat—an area the size of Israel—has been cleared, and most of that has since received at least two follow-up treatments. Surprisingly, there has been very little scientific monitoring of the effects of these activities on water availability. Those studies that have been conducted again show increased streamflows following clearance but have not looked over the long-term. This gap is important because as natural vegetation becomes reestablished it uses more water so that some of the immediate gains from clearance might be expected to diminish over time.

More convincing than the available numbers, however—and perhaps more relevant for much of Working for Water's audience—are the stories of long-dry streams and springs flowing once again. I quiz Edwill and his colleagues for examples. They have plenty. One farmer whose land was cleared of black wattles in the first year of the Kouga project reported that a stream that had run dry in every one of the 40 preceding years had not stopped flowing since. Local springs that had dried up in every drought year over the past half century kept flowing during the exceptional drought of 2003. There have been unexpected gains too. Local apple farmers say that cutting down nonnative trees has increased local airflow to the extent that frost damage has been greatly reduced. Stronger breezes also mean that picking can also start earlier each day because the morning dew is evaporated off the ripe fruits more quickly.

Across South Africa there have been tangible biodiversity gains as well. Clearing pines on the slopes of Table Mountain triggered the germination of large numbers of silver peas—plants recorded nowhere else and whose gray-felted leaves and golden-yellow flowers were until then thought to

8. Working for Water Programme, *The Working for Water Annual Report* 2001/2 (Cape Town, South Africa: Working for Water Programme), p. 16.

have been gone forever. Entomologists searching for long-lost dragon-flies whose only known localities had all been overgrown by alien trees have since rediscovered three of these missing azure jewels; the poetically named Ceres stream-damsel, Cape bluet, and harlequin sprite have all appeared in areas recently cleared by WfW teams. And palmiet, an endemic wonder-plant that cleans up silted water, restores eroded banks, and slows down rising floods, is now recolonizing streams and rivers once choked by black wattle.

Despite its remarkable success, however, Working for Water faces some serious challenges. In terms of its daily operations, restrictions on clearance teams using fire have greatly increased the cost of follow-up operations. On the human side, some argue that the very low wages and the rigid adherence to the poverty relief program's 24-month employment limit its power for social change. They also make for a high turnover rate of workers and hence continuous erosion of the workforce's skill base.

Another worry in the eyes of many observers is the almost complete lack of systematic monitoring of the impacts of Working for Water. As is the case in many environmental programs, almost all reporting has so far focused not on outcomes and impacts—whether streams, people, and nature are better off several years down the line—but on activities, such as areas cleared, people employed, and money spent. That's unsurprising—measuring outcomes and impacts is far harder—but it's rather like trying to monitor the quality of health care from how busy nurses are, when what we're really interested in is whether the patients are getting any better or the population as a whole is healthier. There's been very little documentation of how far the program is helping people into long-term employment, if it's having an effect on overall water availability, or whether (as anticipated) it's doing so at lower cost than the alternatives. Some reckon that these quantitative studies are now largely unnecessary—the case for Working for Water has already been won. But others disagree and worry that as the program ages, ongoing scientific scrutiny will become increasingly important for maintaining political support and large-scale funding.

But these ongoing concerns notwithstanding, Working for Water has been without doubt highly successful. It has created tens of thousands of jobs for some of the poorest people in South Africa, it has made visible inroads into alien infestations right around the country, and perhaps most importantly, it has made clear in the minds of the public and politicians alike that human well-being is intricately linked to the health of the ecosystems we depend on. So why has Working for Water worked?

Winning Ways

As in Assam, one hugely important element in South Africa has been the commitment and determination of a set of core individuals: high-level political champions, a small cadre of visionary scientists, and the program's famously energetic management team. In the early days these people had the foresight to recognize long-term problems and the wit to imagine long-term solutions. Later on, their capacity for plain hard work meant the program could scale up rapidly as poverty relief funding became available. Good science has clearly been pivotal too, in providing the initial evidence of alien trees' thirst, and in demonstrating that clearance generally made good economic sense.

But just as good science rarely follows prescribed lines, perhaps the single most important reason for WfW's success—as for Safe Harbor—has been its proponents' willingness to engage in lateral thought. The ecologists who worried about exotic plants didn't restrict themselves to thinking like ecologists; they thought like people. They didn't just worry about the effect of the problem on their pet species or places but considered the implications for society at large. And while the timing of the switch to democracy was undeniably fortunate, it took clever strategizing by key players to then take full advantage of a novel situation. Broad thinking pervades the everyday activities of Working for Water too—from considering it's an employer's job to teach disadvantaged employees about nutrition and household accounting to seeing cheap coffins in alien deadwood.

Approaching a conservation problem from several angles means many more people come to care. In apartheid South Africa, most citizens had no involvement in decision making. Environmental problems—even those that directly affected people's daily lives—were someone else's responsibility. Working for Water has tried to change that by, as one person explained to me, "giving people who have very little in their lives ownership of both the problem and the solution." By working with local communities rather than ignoring them, Working for Water has turned bystanders and opponents into grassroots champions for the cause.

Broadening the invasive species issue has in turn unlocked high-level political backing, generating not just money but also a strong framework of supporting legislation. The groundbreaking 1998 National Water Act provides a powerful legislative spur for water-saving by granting all South Africans the right to 25 liters of clean water daily, requiring planners to maintain a water reserve for natural ecosystems, and spreading legal responsibility for alien weed clearance to landowners. From the outset, political and popular support has been sustained by the program's high public profile. It has won nearly 40 national and international awards (but makes sure that wherever possible these are collected by ministers). The program's patron in chief is Nelson Mandela. And the bright yellow T-shirts its workers wear are as recognizable as the Nike swoosh.

The Return of the Natives?

What does the future hold? How long WfW will continue to receive support and whether this will be enough to bring South Africa's alien species problem under control remain big questions. Long-term plans talk about a 20-year clearing strategy, but realistically, even 2 decades may not be enough for the Herculean task at hand.

Judging progress to date is hard. Only around one-fifth of the area thought to be affected by alien plants in 1998 has so far been cleared; more detailed calculations adopting the "condensed area" approach (to account for varying densities of infestation) reckon complete clearance of some species could take 80 years. And even these numbers assume that invasives do not continue to spread and that no new species emerge as serious pests.

It seems that getting the upper hand on the unwanted invaders within 20 years or so may require Working for Water to make some difficult strategic decisions. Where should it focus its efforts when the areas of greatest social deprivation do not coincide with those where exotic species are having their greatest effect? Should it risk antagonizing public opinion by putting increased pressure on private and commercial landowners to undertake their legal responsibilities for clearing up? And what about spending more on biological control? This can be good value for money, but employing scientists rather than laborers may be hard to square with poverty relief objectives. Working out how best to resolve all these trade-offs will continue to be difficult.

But although these challenges are real, they're not insurmountable. Complete control of South Africa's problem plants seems unlikely. On the other hand, the significance of making a real dent in their spread while at the same time increasing water flows and providing employment for tens of thousands of people cannot be underestimated.

For me, another Richard Cowling anecdote best summarizes Working for Water's message of hope. Like Norman Moore, the champion of Dorset's disappearing heaths, Richard has despaired at the scale of a problem, fought long and imaginatively to find a solution, and been rewarded by witnessing the first signs of the turning of the tide. Driving through the Kouga catchment a few years ago, he realized that the familiar wattles that had dominated the valley for so long had gone. His beloved fynbos was beginning to come back, in all its exuberant diversity. Rounding a bend, he saw a knot of workers in yellow T-shirts building a weir to restore a wetland. Another team was carrying out follow-up clearance nearby. "Here were people being empowered to make a living out of healing the earth," he says. "Working for Water was simultaneously addressing the two things that had always bothered me most as a South African—social inequality and environmental degradation." Two seemingly insurmountable problems, being tackled together.

REWILDING GOES DUTCH

Working for Water is outstanding but not unique: bold thinking and broad under-standing of the benefits bestowed by nature can lead to dramatic scaling up of con-servation efforts even in the most crowded and highly developed circumstances. A trip around the Netherlands—a country with more economic activity for its area than virtually any other—reveals a nation halfway through an immensely ambi-tious program of handing large swaths of land back to nature. Government moti-vation stems partly from a desire to conserve wildlife, but also, as in South Africa, from a growing recognition that restoring wild places helps people too.

The dusty sun-baked plain is packed with large mammals. In just one sweep of my binoculars I tally more than 600 beasts in half a dozen scattered groups, drifting away from a watering point here; heads down, grazing on the closely cropped sward over there; skittish mothers keeping a watchful eye on a carnivore slinking past to my right. From the vast, cloudless sky, the milling herds, and the pancake flat landscape broken only by the oc-casional emaciated shrub, I can easily imagine I'm still in Africa, on some great grassy wilderness.

Yet in the furthest distance giant wind turbines are turning. A closer look, and the would-be wildebeest morph into red deer, the zebras shed their stripes and become wild ponies, and the buffalo-like creatures loaf-ing in the distance reveal themselves to be primitive cows called Heck cat-tle. The prospective predator is a red fox, scouting for stray goslings that haven't kept up with their crèches. In the middle of the warmest April on record, I am 5 meters below sea level and just half an hour away from Am-sterdam in an extraordinary place called the Oostvaardersplassen. Here, a determined group of biologists have challenged conservation orthodoxy and are rewriting the rules about what nature reserves might look like in twenty-first-century Europe.

Their ideas are influencing an even more audacious experiment. In one of the most densely settled nations on earth, with a larger GDP for its size than almost anywhere else, the Dutch government has committed to giving 17.5 percent of the land back to nature. Building the so-called National Ecological Network—of which the Oostvaardersplassen is just one node— costs around US$1 billion a year and involves reversing decades of painstaking drainage, demolishing industrial estates, and even building motorway bridges for deer.

Nature is edging back into other parts of Western Europe too. Wolves are recolonizing France, brown bears are starting to return to Germany,[1] and lynx have been reintroduced to Switzerland. In the UK, ambitious new programs are restoring large tracts of moorland, oakwood, and Caledonian pine forest. Even where I live, in Britain's agricultural heartland, bold projects are working to re-create functioning areas of the Fens—the watery wilderness that once stretched in a great arc from Lincolnshire through to Suffolk.

But nowhere else in Europe is restoration proceeding on quite such a scale or with such high level support as in the Netherlands. Here, visionary ideas are backed with government blueprints, timelines, and strings of zeros that conservationists elsewhere can only dream of. Over the past 500 years, much of the way the world works has been shaped by Dutch innovations, from merchant banking and stock markets to industrial-scale land reclamation and institutionalized social tolerance. Now, as attention begins to shift from overcoming to rebuilding nature, the Dutch are once again leading the way. I've come to the Netherlands to find out where they're going.

From No Land to Wetland

Heading out toward the Oostvaardersplassen I realize I'm entering a very new part of the world. The unbroken flatness of the immense fields, the utter straightness of the highway, and the absence of almost any trees all confirm that I'm on land that even 40 years ago simply wasn't. Until the 1960s, where I'm driving now was underwater—part of the IJsselmeer, a huge freshwater lake that was itself created some 30 years earlier when engineers blocked off the mouth of what until then had been an almost landlocked sea called the Zuiderzee. Land creation then began apace, with immense dikes built to hold back the water, as the area behind them was steadily pumped

1. . . . . Albeit not always to a rapturous welcome (see chapter 2).

dry to form three enormous polders.[2] Together, two of them make up the largest man-made island on earth. Even now, as the soil dries and shrinks further down, ships' propellers and the remains of planes that crashed into the lake during World War II occasionally emerge in the middle of farmers' fields, ghostly reminders of a very different past.

I drive past the Oostvaardersplassen (which, with remarkable syllabic efficiency, means something like "the place the ships sail through on their way from Amsterdam to the East Indies"), and on to Lelystad, my base for the next few days. Disappointingly, the capital of the new province of Flevoland lives up to its grim guidebook billing as "a good example of urban planning gone awry." I'm sure it has its charms, but around my hotel the gray concrete facades and gridded sprawl remind me of the worst excesses of Britain's 1960s new towns. Luckily, there's somewhere equally new but much more inspiring just 10 minutes down the road: the Oostvaardersplassen Nature Reserve.

Over a drink later that evening, former Oostvaardersplassen warden Vincent Wigbels—neatly bearded and with an infectious smile—explains the history of this unlikely oasis for nature. "On 27 May 1968, the southern Flevoland polder was officially declared dry. But the lowest-lying parts along the edge of the IJsselmeer stayed wet, and the plants and animals began to move in." Vincent describes how the first plants to arrive were ephemeral pioneers like marsh ragwort and golden dock, tough colonists capable of establishing themselves on the bare mud of the newly exposed marsh. Before long their presence opened up opportunities for more long-lasting species like reeds. Birds too colonized the new habitat among them great white egrets (large herons with exquisite breeding plumes that a century earlier saw them heavily hunted to supply Victorian milliners) and spoonbills, also white and heron-like, but with bizarrely flattened bills specialized for catching small fish. Both species were rare in the Netherlands, and both began to breed. A handful of birdwatchers started taking notice.

"The new marshes were only meant to be temporary—the land they occupied was earmarked for large-scale industrial development between the new cities of Lelystad and Almeere. But then the 1973 oil crisis took hold and the Dutch economy fell into recession." Questions were raised about

2. Polders are blocks of now-dry land made by surrounding wetlands, lakes or bits of the sea with high dikes (a.k.a. levees in the southern United States) and then pumping the water out. Keeping them dry has long required the cooperation of many people, even in different cities, so the consensus politics that characterises much Dutch decision making has been labeled the polder model.

the demand for more industry, and financial support for the project melted away. While the money dried up, the marsh—which covered 36 square kilometers, almost half the area of Manhattan—continued to develop into one of the most important wetlands for wildlife in northwest Europe. "An action group took up arms, as did several biologists who didn't want to stand idly by as this beauty disappeared." Mavericks within government helped too.

An embankment was built around the marsh, and—in a complete reversal of the polder paradigm—water was pumped in to keep it from drying out. Somewhat reluctantly, the technocrats slowly abandoned the idea of an industrial park. Planned roads were moved to the edge of the map, and the Oostvaardersplassen was granted the status of temporary nature reserve. This was an extraordinary event in the long history of Dutch polder construction. Never before had so much space been given over to nature. The following day I travel on to the reserve itself to discover what happened next.

Rebels with a Cause

I'm mesmerized by the spoonbill slow-marching across the shallows in front of me. In Britain, spoonbills are still rare birds,[3] and this is the closest I've yet been to one. The clean lines of its head plumes spread out in the light breeze as it sweeps its spoon backward and forward through the water in a hypnotic figure-of-eight. Left and right, left and right. Every few minutes it catches a fish, stops, tosses it up into its mouth. Then back to work, as it walks on again.

Reluctantly, I break off from my spoonbill reverie and walk over to Forest Service headquarters to talk with the three men who've done more than anyone else to shape the Oostvaardersplassen: Frans Vera, tall and surprisingly soft-spoken, given his reputation for controversy; Fred Baerselman, energetic, wiry, and terrier-like; and Leen de Jong, a man with a deep baritone to match his imposing presence. All three were newly appointed biologists in the late 1970s (Frans and Fred in government departments, Leen in an NGO), and as young Turks eager to challenge the system they quickly united around the cause of safeguarding the Oostvaardersplassen. These

3. Once common not far from my home, spoonbills were wiped out 300 years ago. Excitingly, Dutch-born birds are now starting to colonize some of Britain's newly restored wetlands.

days they get back together only rarely, and our meeting quickly resembles a reunion of the Three Musketeers.

We take our chairs outside, and with a cuckoo calling insistently from a nearby poplar, Fred begins. "We started as fresh young ecologists, with lots of new concepts and the will to fight for them. I was particularly interested in island biogeography theory,[4] and what it had to say about the need for nature reserves to be very large." Frans intervenes: "And I was fascinated by ideas about how natural systems change over time. As I found out about what was happening here in the Oostvaardersplassen, I realized that the uniqueness of this place was not in its birds or its plants, but in its openness. This place could be a unique experiment into ecological processes. But none of the officials in charge of conservation wanted to know."

"For a long time," according to Fred, "we were told that thinking about the Oostvaardersplassen was not our job. But we disagreed—and we learned to use the system." Reminiscing turns to laughter as Frans recalls how he was told to stop writing a report on the problems facing the area. "But then Fred got his bosses to commission a report from my ministry, which only I could write." Leen joins in: "We pulled off lots of tricks like this. We got leaks about what was happening from friends in other departments, so between us we always knew more than our bosses did. We really enjoyed these games." "Above all," concludes Fred, smiling wryly, "we had a great deal of fun."

At the end of the seventies, several problems exercised this subversive triumvirate. Much of the new southern Flevoland polder was being taken over by intensive agriculture, industry, and housing, so that many of the species that foraged beyond the reserve were having a tough time. This underscored the importance of maintaining the integrity of the Oostvaardersplassen itself. Here, however, a border zone of about 20 square kilometers of drier, grassy ground that buffered the reserve from the surrounding fields and helped maintain the ecology of the wetland was being developed for farming and looked like becoming separated from the marshes by a new railway line from Amsterdam to Lelystad. The reason this drier area mattered to the survival of the wetland lay in the ecological processes that so fascinated Frans.

Habitats rarely stand still. Over years and decades they exhibit eco-

4. A theory developed in the 1960s by the great U.S. ecologists Robert MacArthur and E. O. Wilson explains why small and distant islands have relatively few species; Jared Diamond later adapted it in order to devise guidelines for nature reserve design.

logical succession, with open communities becoming progressively more closed over time until they're opened up again by a disturbance event (like a fire or a flood or a landslide) that resets the clock. Without this sort of interference, grasslands typically become open woodlands, open woodlands (like the longleaf forest of the Carolinas) grow thicker, wet marshes dry out into scrub, and so on. Yet to many people's surprise, this didn't happen to the new reedbeds of the Oostvaardersplassen. For the most part they stayed relatively open, held back in an early stage of succession that suited many of the animals and plants that made the area so special. What was responsible for Oostvaardersplassen's state of suspended animation? Greylag geese, and those dry grasslands around the edge of the reserve.

Greylags are gray-brown, stocky birds that resemble the geese of nursery rhymes.[5] Soon after the creation of the Oostvaardersplassen, tens of thousands of greylags from around Europe started converging on the wetlands each May, using the area as a safe haven for their annual molt. For the month it takes to grow new wing feathers the geese are incapable of flight, so they spent most of this time hiding deep among the reeds, feeding on their leaves and underground stems. This onslaught of grazing and digging was enough to stop succession in its tracks: the reeds couldn't take over completely, and the wetland stayed wet.

The problem that Frans spotted was that for a week or so either side of molting, the geese fed out on the adjacent grasslands of the border zone. If this area was converted to farming or got cut off from the marshes, the geese might end up choosing somewhere else to molt, and succession would set in - so without the grasslands, the wetlands could dry up. To avoid this, Frans, Leen, and Fred campaigned for the precious grassland of the border zone to become an integral part of the reserve.

They also argued that the best way to produce the short, nutritious swards that the geese prefer was to introduce large mammals similar to those found in European grasslands in the past: red deer, a wild cow called the aurochs, and a wild horse known as the tarpan. There was a slight problem here, in that the last two species were already extinct (dying out in 1627 and 1887, respectively). But Fred and Frans—not the type to be put off by small details like extinction—suggested that herds of the more primitive breeds of domestic cows and horses would probably have much the same impact.

Remarkably, those battling for the Oostvaardersplassen eventually won

---

5. Not surprisingly, given that they are the ancestors of the domestic goose—Mother Goose's mother, in effect.

again. Through a combination of hard work, skulduggery, and sheer native wit (and thanks in no small part to a birdwatching government minister who backed the rebels over his civil servants), Vera, deJong, and Baerselman got their way. After two or three tumultuous years at the start of the eighties, the reserve was extended to include nearly all of the border zone. The planned railway line was rerouted. And Fred and Frans were given permission to go and look for extinct horses and cows.

Of Mowers and Men

The Heck cattle that they brought back to the Oostvaardersplassen may be smaller than the mighty aurochs celebrated in the cave paintings of Lascaux, but they're impressive beasts nonetheless. The black or chocolate-brown bulls weigh in at nearly a tonne, their daunting bulk relieved only by incongruous mops of blonde hair that sprout between handlebar horns almost a meter across. Females are smaller and browner and spend most of the year away from the males, tending their golden-red calves in nursery herds. Despite its primitive appearance the breed was created in the 1920s, when two German brothers, Heinz and Lutz Heck, started crossing modern varieties of domesticated cattle in an attempt to recreate the aurochs. While their goal was always genetically unachievable (as the particular combinations of genes that code for an aurochs were long gone), the eponymous breed the Heck brothers produced is better suited than most for surviving with very little human support: since the early eighties, numbers at the Oostvaardersplassen have risen tenfold to over 300.

The other large grazers that Frans and Fred brought in have fared even better. Red deer—which until then had been missing from the rich feeding grounds of the Dutch lowlands for decades—now number over 2,000 and spill across the plains of the border zone in every direction. In amongst them are over 600 koniks—primitive ponies that are about the closest living things to tarpans. Stocky, fawn-gray, and with lax black manes, koniks were the product of seventeenth-century Polish farmers' attempts to strengthen their domestic horse bloodlines by cross-breeding them with the last few wild tarpan. At Oostvaardersplassen the koniks have proved even tougher than the Heck cattle.

Collectively this army of grazers has certainly kept the buffer zone grasslands short and attractive to geese, but other interventions have been necessary too. Because the area used to be the seabed and so is as flat as a pancake, managers had to dig out a complex system of pools and channels to provide habitats for fish, frogs and aquatic insects, and hence feeding grounds for high-priority birds like spoonbills and egrets. Uniformity was also becoming a problem in the marshland, where grazing by the greylags was so effective that the reedbed was in danger of disappearing completely in their web-footed wake. The solution was to divide the marsh in two and drain one section. While the geese and other birds concentrated on the wetter side, the reeds recolonized the freshly exposed mud on the drier portion. A few years later, this was flooded again, and the goose pressure was dispersed across an enlarged area of reedbed.

All of these approaches to managing the Oostvaardersplassen have been guided by three underlying principles that potentially have relevance right across the continent but that remain almost as controversial as they were when Frans, Fred, and Leen initially formulated them. The first is a belief that the prehuman habitats of Europe were much more open than generally thought. Paleoecologists—scientists studying past environments—get insights into the former vegetation of a place by looking at changes in the makeup of pollen trapped in the sediments of small wetlands, often tens of thousands of years ago. In Europe, much of this ancient pollen comes from trees (rather than the grass pollen that predominates in most places nowadays), supporting the conventional view that, before clearance by people, most of the region was covered by so-called wildwood (rather than more open habitats).

But Frans has pointed out that pollen of two of the most commonly represented tree species—hazel and oak—is actually indicative of open conditions: hazel flowers only in clearings, and acorns grow into oak trees only away from the gloom of the forest canopy. Add the pollen of these two species to those of the grasses, he argues, and suddenly many areas appear in the past to have been predominantly open—more like savannahs and less like closed forests. Vera believes this was in part because large herbivores (like aurochs, deer, and tarpan) were once common enough to drive a cycle of succession, combining with floods, fires, and storms to occasionally open up closed woodlands and convert them into grasslands. Heavy grazing pressure did something else too: it favored grazing-resistant thornbushes, which in due course acted as impenetrable nurseries where young trees could germinate. Protected from grazing by the tough bushes around them, vulnerable saplings could escape from herbivores long enough to mature into adults and form new woodlands. But in the fullness of time, these would grow old and be opened up by grazers once again. Large herbivores played a key role at each stage in the cycle, hence the central importance to Vera's vision of ecological restoration of now reinstating these outsized mowing machines.

The second principle behind the Oostvaardersplassen experiment is that we need to look further back in time in setting goals and standards for conservation. The wildest ambitions of conservationists are often not ambitious enough because they adopt as a baseline whatever is familiar to them—yet we began wearing away at nature very long ago, so a contemporary baseline may often be a highly degraded shadow of an ecosystem's former glory.[6] Nowhere is this more the case, Frans and his colleagues argue, than in northwestern Europe. As early farmers cleared the region's natural vegetation and steadily replaced it with agriculture, many species (such as wolves, lynx, bears, and bison) were lost from all but the most remote areas, or (like the aurochs and tarpan) went extinct altogether. Other species hung on in small habitat patches like hedgerows, copses, or farm ponds, while some (such as skylarks, lapwings, corncockles, and poppies) adapted to the novel farmed landscape. They became to many Europeans the archetypal creatures of their countryside. But more recently, with the rapid intensification of farming following World War II, these species too began to decline, unable to cope with the loss of much remaining habitat, the onslaught of pesticides and fertilizers, and the increasing uniformity of ever-more efficient agriculture.

In response, Europe's conservationists have bought up remaining woodlands, meadows, and wetlands. Because these fragments have been too small to sustain natural ecological processes, managers have then had to imitate them—bringing in sheep to keep chalk grasslands just the right length for rare butterflies, maintaining woodland flowers like violets and primroses by reinstating traditional coppicing, and using small armies of volunteers to cut back encroaching scrub from pocket-handkerchief reedbeds. Most recently, conservation organizations have also targeted farmers, persuading them to soften some of their practices and make their farms friendly to wildlife once more.

But while these interventions are helping some of the most badly hit countryside species to recover, Frans and his colleagues reckon such efforts are fundamentally misdirected. The wildlife we're striving so hard

---

6. This idea of so-called shifting baselines was first put forward by marine biologists to describe the tendency of fisheries managers to overlook historical data and instead to try to manage fish stocks at the level they encountered at the start of their careers. As populations decline this inexorable ratcheting down of expectations keeps even well-managed populations far below those that could be maintained under more ambitious management.

to retain is mostly that of man-made, cultural landscapes, when what we should be doing, they argue, is fighting to bring back natural landscapes and the ecological processes that drive them. Ironically, this more ambitious goal may be made more achievable by agricultural intensification. By enabling demand for food production to be met on a smaller total area of land, intensive farming has the potential to free up large areas elsewhere for nature—and that's exactly what nature needs. Really big wetlands, fully functioning floodplains, seriously large forests—and all the species that once inhabited them—these are what conservationists should be focusing on. For Vera and company, the way forward, even in crowded countries like the Netherlands, is restoration on a truly massive scale. Anything else is missing the point.

The third tenet of the Dutch radicals is that both practically and as a point of principle, the only way to manage such large new areas of nature is to reintroduce the ecological processes that shaped them in the past—abandon our own attempts to mimic nature by reed-cutting and coppicing, making artificial nest holes, and digging ersatz ponds, and instead reinstate natural grazing and flooding, decay and succession. Alongside that, they argue, conservationists need to let go of numerical targets. Rather than aiming to have x pairs of lesser-spotted woodpeckers or y hectares of dry acid grassland, perhaps to the exclusion of other important species or habitats, conservation should seek to restore functioning ecosystems, within which nature and not people will determine what lives where, and how that changes over time.

Keeping the vast Oostvaardersplassen wetlands from maturing into scrub and eventually turning into woodland would have been impossible with just a man and a mower. Instead, the twin engines of the molting greylags and thousands of large grazers keep the system open without the need for intensive human intervention. But Frans, Fred, and Leen go further, arguing that pretty much all interventions are undesirable. Under this hands-off approach, a decade ago even the alternating draining-down and rewetting of the marshland was abandoned. The Oostvaardersplassen is now on automatic pilot.

Challenging Unorthodoxy

Applying these three principles—encouraging openness, thinking big, and striving for self-determining processes rather than hands-on targets—has yielded some impressive achievements. Over 250 bird species have been recorded in the Oostvaardersplassen. More than 100 of these breed, and for nearly 30 species the reserve is home for at least some of the year to 1 percent or more of their entire western European populations. Up to 10,000 pairs of cormorants nest there. As many as 12,000 smew—dapper black-and-white ducks that breed in the forests of the Russian Far North—overwinter on the Oostvaardersplassen's lakes. And after a mild winter, up to 50 pairs of bitterns set up territories deep in the reedbeds.

For Vera, most exciting of all has been the return of one of Europe's largest birds of prey—the white-tailed eagle. Closely related to North America's bald eagle, this enormous raptor had long since disappeared from the Netherlands, but as early as 1980 he predicted that the Oostvaardersplassen had enough food, space, and peace and quiet to host one or two pairs. The birds started renesting in 2006, triggering nationwide excitement. Over 15,000 people now log on to the Forest Service's webcam each day to see how the latest chicks are doing. Other new arrivals are anticipated, with hopes of osprey moving in, and in due course of reintroducing Eurasian beaver, wild boar, moose,[7] and maybe even bison. Yet despite these successes, many people are concerned about the future of the Oostvaardersplassen and about the radical ideas that have shaped its development.

Unsurprisingly, many paleoecologists have challenged Frans's reassessment of past vegetation. Some argue that their original interpretation of the signal from the pollen deposits was correct. Others suggest that even if there was cyclical succession, most of the prehuman landscape would, at any one time, have been closed. Many concede that there may have been more gaps in the wildwood than originally thought, but even then, they argue, this may have been due more to fires and flooding than to grazing. In short, they question whether this part of Europe ever had African-style savannahs, and therefore whether grazer-driven openness is a worthwhile conservation goal.

There's a much more public controversy about big grazing animals too—one that concerns the way they die. To make the grazing regime as natural as possible the Oostvaardersplassen herbivores aren't fed by people (except in extremely harsh cold spells), so their numbers are limited by how much

---

7. Confusingly, the enormous semiaquatic deer that North Americans call moose (collisions with which are responsible for the sturdiness of Volvos) are referred to by anglophone Europeans as elk, while Americans use the name elk (or wapiti) for the North American relatives of what Europeans call red deer. Meanwhile, the huge wild cattle that Europeans refer to as bison are called buffalo in North America, a name Europeans use only for the African buffalo and the Asian water buffalo. To try to limit the confusion here I'll do my best to call all moose "moose," all bison "bison," North American elk "elk," and European red deer "red deer."

food they can find for themselves. By the end of winter many are thin, and in bad years up to a third die. This upsets people concerned about animal welfare, particularly when the animals involved are cows and horses. The Forest Service's response is to limit suffering by shooting animals that are obviously starving. Some people argue a preemptive cull before each winter would be kinder, but Frans points out that because predicting over-winter survival is tricky, this would be undesirably arbitrary—individuals that might otherwise survive would be shot, and there would be less scope for natural selection to weed out animals less well adapted to fending for themselves. He also questions the welfare lobby's preoccupation with how Oostvaardersplassen's cattle and horses die, arguing that the social freedoms they enjoy while alive—keeping their calves with them, choosing their mates, and so on—are at least as important. They die thin, but they live happy.

A dramatically different take on the death debate is the suggestion that what Oostvaardersplassen really needs are wolves. Big predators might not only do the Forest Service's job of speedily dispatching the old and infirm: they might also lower herbivore numbers so fewer animals died of starvation. It turns out, rather surprisingly, that evidence supporting these arguments is somewhat equivocal. In the Serengeti, for instance, lions, hyenas, and leopards kill a far lower proportion of starving animals than the Oostvaardersplassen rangers do. And despite facing a fearsome array of predators, population sizes of Africa's bigger-bodied grazers still seem to be mostly controlled from the bottom up (i.e., by how much they get to eat) rather than the top down (by how much they get eaten).

On the other hand, fear of predation does shift the distribution of prey animals away from areas with good forage toward places of safety. And as the recent return of wolves to Yellowstone (mentioned in chapter 1) is beginning to reveal, this redistribution can have major knock-on benefits. Nevertheless, while restoring top-level predators to the Oostvaardersplassen and beyond might be a desirable long-term goal, most conservationists reckon it's a while away yet. Wolves require more space than is currently available, and there's too much work to do in the PR department.

Hungry herbivores and fear-inducing wolves aside, there are other, more general questions about where the Oostvaardersplassen might be heading. Reinstating natural processes is widely accepted as making good sense, and a target-free, noninterventionist approach may be laudable where a system is sufficiently pristine, large, and isolated to be immune to the effects of people. But critics argue that the Oostvaardersplassen is none of these things. Its unnatural origins and resulting flatness and the absence of any rivers running through and periodically flooding it together mean that without repeated human interference—intermittently changing water levels in parts of the reserve, for instance—its habitats remain vulnerable to sliding into unhelpful uniformity.

There are concerns about grazing levels too. Civil servant and long-term champion of the Oostvaardersplassen Frank Alberts thinks that herbivore densities have become too high. "In really large, unfenced areas some places get left ungrazed for a few years, and recover. But at Oostvaardersplassen the grazers have nowhere else to go, and so the constantly high herbivore numbers may simply be too much for other animals." Some fear that the whole of the buffer zone is now turning into short grassland. Birds that depend on longer vegetation and shrubs—like stonechats and corncrakes—are declining. So too are voles (which can't find enough seeds to eat when the grass doesn't get tall enough to flower), and spectacular vole-eating raptors like hen harriers and rough-legged buzzards—all far less numerous at Oostvaardersplassen now than 20 years ago.

Losing some species like this seems to be an inevitable consequence of focusing exclusively on ecological processes and no longer managing for species-based targets. Vera and his colleagues argue that this is not a bad thing. Provided conservation areas are big enough and well enough connected to one another, species can disperse elsewhere, so a downturn in conditions locally needn't mean population extinction. Others worry that at the moment the Oostvaardersplassen may still be too small and too isolated to cope on its own. Without active intervention, valuable species drop out, and it's not always clear they have anywhere else to go.

So there are real disagreements on how much openness is a good thing, the ethics of big furry animals starving, and laissez-faire versus target-driven management. But where Dutch opinions converge is over the groundbreaking importance of the other ideas that the Oostvaardersplassen experiment has thrown up—about the scale of conservation that might be possible, about restoring processes, and about the essential role of connectedness: ideas that are at the core of the country's even bolder National Ecological Network.

Bridging Gaps

To see how the network is taking shape on the ground, I've come to the edge of the town of Hilversum, where I'm struggling to absorb the enormity and sheer strangeness of what I'm looking at. Three massive piles of earth with young trees on their flanks each rise up as high as houses. They're connected in a long line by a couple of enormous concrete plinths, themselves topped with more soil, bushes, and even saplings. The vast structure is an ecoduct—a new kind of bridge built so that animals can cross safely between two previously isolated patches of habitat. Opened by the Dutch queen just 12 months before my visit, it's one of the latest pieces to be completed in the Netherlands' grand master plan for restoring nature countrywide.

It's not just any ecoduct, either. As Rob Rossel, a ranger with local conservation NGO Goois Natuurreservaat and one of the people who built this bridge proudly explains, "Crailo is the largest ecoduct in the world. It's over 50 meters wide, and it stretches for nearly a kilometer between Spanders Wood on one side and a heathland called the Westerheide on the other." Along the way it crosses a main road, the Utrecht-Amsterdam railway line, a rail-manufacturing plant, and a sports complex. Four hundred trains and 16,000 cars pass beneath the bridge every day. It's a busy place.

The aim is to have a lot of traffic on top too—but of a very different sort. Rob and his team have worked hard to create a new piece of habitat that animals might wander into and then walk across. They built it 50 meters wide based on experience in Switzerland, where big animals have proved reluctant to use narrower bridges. Rob shows me how they've landscaped the ecoduct to resemble a hill. The colder, north-facing slopes are planted with hawthorn, hazel, blackthorn, and birch, while gorse and heathers are growing on the sunnier southern banks. Along the top of the bridge, piles of deadwood and even ponds have been put in as stepping stones to entice smaller creatures. Discreetly screened fences stop animals from falling off or trying to cross the railway tracks. The new vegetation helps dampen noise from the traffic below and shields capping the railway's floodlights keep the area in relative darkness at night.

The reason for going to all this trouble is to try to reverse the effects of habitat fragmentation. Under natural circumstances, different patches of habitat and the animals and plants that live in them are not usually isolated from one another. Some of their inhabitants will range widely, crossing between patches. In more sedentary species, young animals will nevertheless often disperse away from their natal populations and set up home elsewhere. Plants too will move around: pollen will be carried by animals and blown on the wind, and seeds will be transported in animals' guts and on their fur. All this movement means plant and animal populations remain viable over long periods. Individuals with genes coding for favorable adaptations will occasionally move into and out of populations. Species can recolonize areas if populations die out. And crucially, if environmental conditions (such as climate) change, populations have a chance to shift their distributions in response.

But over the past century not only have we reduced the size of habitat patches, we've broken many of the links between them too. We've built railway lines, canals, roads, and then more roads, all of which make life extremely hard for anything that can't fly (and even quite a lot that can). As a result, populations have become not just reduced but also cut off. Those that are too small to persist without replenishment are dying out. Barriers to dispersal in turn mean that lost populations are not being easily replaced. And as human-induced climate change begins to take hold, more sensitive and less mobile species are finding it difficult keep up, unable to move into new climatically suitable areas fast enough. Nowhere is this more the case than in northwestern Europe—a region with extreme habitat fragmentation and more tarmac roads per square kilometer than just about anywhere else on earth.

Faced with trying to create a future for wildlife amidst a network of raillines and roads crisscrossing the country like the threads of spider web, Dutch planners have opted for ecoducts as part of the solution. The Crailo nature bridge is one of half a dozen built or planned around the town of Hilversum alone. If you have the chance, zoom in on the Netherlands using Google Earth and you can find others strung out along the main A1 road from Amsterdam to Germany. There's a particularly clear one linking two woodlands southeast of Oldenzaal, just to the west of the German border. The National Ecological Network may eventually include 20 or more nationwide.

And there is evidence that ecoducts are starting to work. Infrared cameras, surveys of droppings, and sand strips laid to capture footprints are together turning up signs of frogs, mice, lizards, and foxes all using the new bridges. The Crailo ecoduct—designed for a whole suite of animals, from grass snakes to rare sand lizards and badgers to beetles—was seen being used by one bold roe deer even before the bulldozers had finished. It seems as though ecoducts may indeed help animals disperse in and out of habitats that have become cut off by human infrastructure. Progress doesn't come cheap—Crailo took over 4 years and nearly €15 million to build. But as Rossel points out, in a country where it can cost €5 million to build one roundabout, it may be a price worth paying for helping nature cope with the traffic.

The boldness and imagination that's needed to build multimillion-dollar motorway bridges for wildlife is typical of the Dutch National Ecological Network as a whole. First conceived in the late 1980s in response to

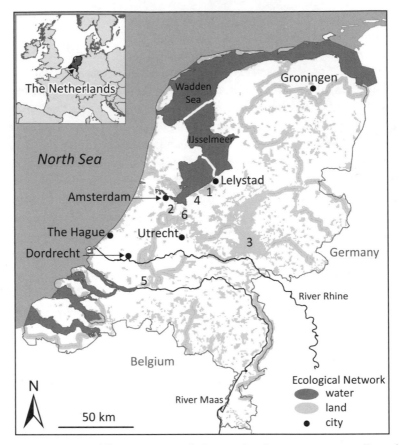

Map 5. The 56-square-kilometer, man-made Oostvaardersplassen nature reserve lies at the center of an ambitious program of conserving, creating, and connecting natural habitats right across the Netherlands. 1, Oostvaardersplassen; 2, Naardermeer; 3, Hoge Veluwe National Park; 4, Horsterwold; 5, Biesbosch; 6, Hilversum. Map based on information in Hootsmans and Kampf 2004. Cartography by Ruth Swetnam.

the realization that the area of the country under nature had halved since the turn of the century, the network aims to create 3,000 square kilometers of new wildlife habitat (some 7 percent of the Netherlands' land surface) by 2018. Relatively speaking, that's like the United States deciding to convert the whole of Texas into a nature reserve. Added to existing natural habitats this means that eventually over one-sixth of Dutch land will be managed primarily for conservation. On top of that, almost all of the country's estuaries, as well as the IJsselmeer and the Wadden Sea, are included in the network too. In proportional terms, no other country in Europe is aiming to restore nature on such an enormous scale.

The overall budget for the network is staggering. The annual allocation from national government is now running close to US$500 million a year. Additional inputs from local government, the European Union, a national lottery, and the rank-and-file membership of conservation NGOs together come to a similar amount. The extraordinary scale of the investment reflects something even more important—the motivation behind it. Not just a concern for wildlife but a broad understanding, right up to the highest levels of government, of the contribution that nature makes to human well-being.

Politicians recognize that the National Ecological Network will benefit not just other species but people too. New wetlands will store floodwater and so safeguard property and businesses. They will help mop up pollutants as well. Growing forests will turn carbon dioxide into trees and so help offset the country's emissions of greenhouse gases. And more green spaces will help meet rapidly rising demands for recreation as well as for simple peace and quiet. The government realizes that because many Dutch people nowadays work under psychological rather than physical pressure, places that offer relaxation and tranquility are becoming ever-more important.

It's little over halfway into the time allocated to build the network, and there's been substantial progress. Rough national plans have been passed on to provincial and community authorities to be translated into detailed blueprints, and serious amounts of money have followed. Around 1,000 square kilometers of former farmland has been turned into new areas for nature. Some of it has been purchased outright, but an increasing proportion has been brought under conservation management through agreements with farmers and other private landowners. In line with the big-is-beautiful thinking behind the Oostvaardersplassen, many of the newly designated areas have been picked to help enlarge existing blocks of habitat. Yet fragmentation has proceeded so far in the Netherlands that many pieces of the jigsaw remain small and isolated.

To help tackle this, major corridors will connect the core areas and link them to many of the smaller reserves along the way. Those charged with designing these connections—people like Hans Kampf from the Ministry of Agriculture, Nature, and Food—developed a vision of a red deer one day being able to walk from the Oostvaardersplassen all the way to Germany without risking its neck by crossing a road. Some key stepping stones are in place, most significantly, the still-extensive forests and heathlands of the Veluwe, southeast of Oostvaardersplassen. Joining up these habitat islands—"de-fragmentation," as Hans calls it—will take enormous efforts. But as he explains, while we pore over the colored sketchbooks produced to bring the corridor concept to life for decision-makers, "Dreams can become reality if you really want them to. It all depends on being able to enthuse people about your ideas."

The ecoducts at Crailo and around Hilversum are some of the first results of this push for connectedness. They're links in a chain that will eventually join the important wetland of Naardermeer, near Amsterdam, to the Veluwe, 50 kilometers to the southeast. Sitting in the provincial government offices in Lelystad, I find out about plans to plug the Oostvaardersplassen into the national network too. The province of Flevoland is buying up farms in a 10-kilometer by 2-kilometer strip running between the Oostvaardersplassen and the Horsterwold Forest, halfway to the Veluwe. The geometrically precise fields of tulips and onions will be uprooted and replaced by a patchwork of marshes and woodland. The program manager tells me this part of the corridor will be ready by 2015, linking together a contiguous block of habitat covering more than 100 square kilometers. The budget for this work alone is over US$300 million. "But that's not so much," says the manager, with remarkable nonchalance. "Not when you look at the cost of building highways for people."

Besides bridges and tracts of new habitat there's another crucial component to stitching the network together, one that more than any other illustrates the potentially powerful alignment of conservation and human self-interest. It stems from a series of national emergencies in the 1990s and it focuses on restoring nature's own great corridors: the rivers.

Eager Beavers

Over half of the Netherlands lies below sea level—a reality that for centuries has concentrated Dutch people's minds on the risks of flooding. Historically, the threat has largely come from the sea, and the response has been an elaborate network of fiercely defended dikes that protect the country from storm surges and high tides. But in January and February 1995, the floodwaters came instead from the rivers, as extremely high rainfall in Belgium and a sudden snowmelt in the Alps caused the Rhine and the Maas (or Meuse) to burst their banks. A quarter of a million people were evacuated, four died, and several cities were flooded; only high river dikes prevented much greater damage. Similar events just 14 months earlier had been put down as a once-in-a-century occurrence, but the 1995 floods forced a rethink. With climate change predicted to result in heavier and more concentrated winter rainfall, and with more and more infrastructure packed into vulnerable low-lying areas immediately behind the river dikes, the government opted for a radical change in the nation's flood defense system.

Traditionally, Dutch river defenses have consisted of tall winter dikes (built to contain all floods), smaller summer dikes immediately alongside the river, and between them, low-lying forelands that absorb floodwaters in winter but are used for farming in summer. Nowadays, however, the long-term build-up of sediment on the riverbed and forelands and the increases in peak flows mean the system is at breaking point. There's much more water and much less space to hold it before it gets to the sea. In response, the government has come up with Room for the River, a 10-year program designed to reduce the risk of severe floods (which breach the winter dikes) by allowing rivers to flood more naturally. In key areas along each of the major rivers summer dikes will be taken down, farms removed from the forelands so that they can store more water, and river-beds deepened. As a byproduct, floodplain habitats will be expanded and the natural dynamics of the rivers partly restored. Revitalized rivers will become important corridors in the National Ecological Network. Making room for the river will make more space for nature too.

To find out firsthand what making room for a river means, for the final leg of my journey I've traveled to the southwest of the country to a magical mosaic of islands and river channels called the Biesbosch. Lying in the heart of the delta where the Rhine and the Maas fan out to meet to the North Sea, the Biesbosch is a special place, and as a naturalist with a near-pathological interest in seeing weird and wonderful beasts I confess I have an ulterior motive for coming here: it's one of the best sites in Western Europe for seeing Eurasian beavers. As a conservationist, my reason for visiting is that the Biesbosch area is also the focus of a couple of enormous Room for the River projects.

The projects are hard to miss. As I drive out along the Biesbosch peninsula I'm once again taken back by the scale of the work being undertaken. Carefully ordered fields abruptly give way to what looks like an enormous construction site, where giant earth-moving machines are loading up equally vast trucks with tonne after tonne of soil. Houses, farmyards, whole farms have disappeared. An entire polder is being removed, cubic meter by cubic meter, and the earth used to create what will become islands in a new wetland. In due course the dike will be breached, and the river will spill out onto what used to be farmland, finding its own new course toward the sea.

It's a vast engineering (or perhaps de-engineering) operation, covering 6 square kilometers. And this is the smaller of the two projects. The second phase, which will take another 8 years and US$500 million to complete, will be four times as big. The combined benefits: a floodplain with much more water storage capacity, a 35-centimeter reduction in the peak level of the river, and a greatly reduced risk of flooding for the inhabitants of the city of Dordrecht a few kilometers to the north. Oh, and quite a lot more space for beavers.

Like their darker-furred North American cousins, Eurasian beavers have for the most part had a pretty tough time at the hands of people. They once ranged from France right across to China. Yet by the start of the twentieth century there were only an estimated 1,200 left, victims of overhunting for their fantastically valuable fur, a secretion they produce called castoreum (used in perfumery and medicine), and their meat.[8] The last Dutch beaver was dispatched in 1826, but a reintroduction program using East German animals began in 1988. After a slow start, the animals bred, and there are now over 200 homegrown beavers paddling around the Biesbosch.

To learn more about these elusive mammals, I spend a balmy afternoon pottering through winding creeks and quiet marshes in the company of park ranger and beaver doyen Dirk Fey. He's worked in the Biesbosch for over 40 years and looks like you'd imagine Sean Connery would if he'd spent that long out in the field. He points out the signs of the beaver population he's helped reestablish with pride in a job well done: freshly felled alders with neatly chiseled stumps fringed by wood shavings; ridged toothmarks where nutritious bark has been gnawed off; and branches cut short, their juicier twigs taken away for storage in underwater larders. Dirk noses the boat past a bank lined by rodent-coppiced willows, where regular haulouts have cut notches deep into the woodland beyond. Few other animals besides people have such evident physical impacts on their environment.

We get out at one cutting and walk to a damp spot where several trails intersect and three different beaver families leave their scent. Dirk grabs a handful of mud and inhales. To him it smells like single malt. Perhaps he has spent too long in the field—for me it's more like antiseptic. We move on and find a lodge. Much smaller than a North American beaver home, this one's a pile of logs and mud around two meters high and twice as wide, deeply embedded in the bank of a quiet channel. As we tiptoe around, the invisible occupants are just the other side of the earth-and-wood wall, so close we can smell them. Tantalizing.

But beavers are decidedly nocturnal, so to have a chance of seeing one

8. Beavers' aquatic habits and flattened tails meant they were classified by the Roman Catholic Church as fish and so could be eaten on Fridays.

in the flesh I book an evening beaver safari. Twenty other people have the same idea. As dusk descends, we cruise slowly along tranquil backwaters, scanning the surface for movements. Catkins drift gently past, the chill evening air rises off the water, and we keep peering. An hour on into the fading light and our guide spots a telltale ripple. We inch forward, hold our collective breath, and wait. And then it rises to the surface, almost in touching distance. It's the size of a Labrador, with wet-spiked fur, small ears, and a great black paddle of a tail. A small, glinting eye takes in the boat-bound visitors for a minute or more. Then, curiosity satisfied, the beaver slips down and is gone. We all smile, happy witnesses to the return of the native. As Dirk says, "It's a nice animal. People like it."

Rewilding Writ Large?

After a day like that it's hard not to draw clichéd analogies between the metaphorical (and literal) industriousness of the beavers and the ambition and energy of the Dutch attempts to restore their wildlife and wild spaces. Ecoducts, corridors, de-fragmentation, even dike dismantling. I've lost counts of the dozens of projects, the thousands of hectares, and the billions of dollars. But the big questions are clear enough: will it all happen, and will it work?

The answer to the first seems largely yes, but with pretty much unavoidable delays and caveats. Just about everyone I talk with reckons that realistically the network won't be finished by 2018, and that there are still some important barriers to progress. Inevitably, land acquisition has so far focused on cheaper and less controversial areas. The low-hanging fruit has already been picked. The remaining gaps in the system are often in places where there are competing economic interests and where conservation will be correspondingly costly, both financially and politically. There are concerns too about the long-term prospects for those areas brought into the network through agreements with farmers and other private landowners, rather than bought outright. In the short-term these save money, but there are fears that the cash to fund them will run out, so will they continue to deliver conservation benefits in 20 or 30 years' time?

These worries are undoubtedly serious, but despite them it seems clear that the network will continue to grow and strengthen, for the simple reason that people want it. It may cost more around US$1 billion each year, but as Hans argues persuasively, that works out at less than half the cost of a stamp per citizen per day. The Dutch public and their politicians have evidently decided that's a price worth paying for bringing nature back and restoring the benefits they get from it.

Will the network manage to achieve its aims? In broad terms, it is starting to. The downward trend in the area occupied by natural habitats is being reversed for the first time in over a century. As with Working for Water, many argue there is a need for broader, more detailed monitoring, to understand which specific interventions are working (or not), and why. There are also important unresolved debates (as at Oostvaardersplassen) over how far the network should focus on quantitative targets for populations and habitat types versus restoring ecological processes and then letting nature take its course. And there's important food for thought about the enormous cost of large-scale restoration, its probable reliance on having areas of intensive agricultural production elsewhere, and the simple point that conservation would be a lot less expensive if we didn't lose wild places and populations in the first place.

But observers agree that the Dutch emphasis on a big network, individually large areas, connections between them, and the pluses for people as well as wildlife together constitute a major step forward in European conservation thinking. As I explore in the next chapter, this new focus on the benefits people get from nature is driving conservation forward in many parts of the world. And the idea of large-scale networks for nature is catching on too. The Dutch plans now dovetail with similar blueprints for Germany and Belgium. As I write this, Europe's environment ministers are getting together to agree on an entire pan-European network. If this were ever realized, the Oostvaardersplassen's roving red deer needn't stop in Germany—it could carry on trekking all the way to the Urals.

The overwhelming message I take from this and from my time in the Netherlands is that the shifting baseline and the unidirectional loss of nature since the industrial revolution could just conceivably be reversed. At least in well-fed countries, the fundamental constraints may not be money or space, but imagination and aspiration. If conservationists are bold enough and if politicians are brave enough, we can begin to challenge the endless downgrading over time of our expectations for the state of the natural world. Future generations might reasonably expect their grandchildren to live in a world with more and not less wild spaces, with missing species restored rather than gone forever. As the ever-energetic Fred Baerselman put it at the end of our chat in the Oostvaardersplassen, "if you can do it in Holland, with all its people, and all its infrastructure, you really can do it anywhere." If he's right—and I sincerely hope he is—who knows what Europe will look like in 100 years' time?

SEEING THE GOOD
FROM THE TREES

Recognizing the benefits that people get from maintaining or restoring intact eco-systems is an idea that is inspiring conservation far beyond South Africa and the Netherlands—it's reshaping the conservation landscape worldwide. This chapter tells to the stories of three groundbreaking initiatives, each based around valuing these so-called ecosystem services. Though very different from one another they all involve seeing habitats in fresh ways and as a result building powerful new partnerships for slowing nature's decline.

Don Alejandro Ramírez smiles warmly as he gestures for me to take a seat in the modest, concrete-floored front room of his one-story house. It's a few days before Christmas, and we're in Loma Alta, a small village on a dusty road winding from the Pacific coast up into the foothills of western Ecuador's Colonche Mountains. Brown-skinned and with jet-black hair that belies his 70 years, Don Alejandro is one of the most respected elders of his community, and a man whose piercing eyes have witnessed dramatic change. I follow the bare bones of his quietly told story in my rusty Spanish, and Tanner (an American volunteering in the village) fills in the details.

"When I was 14 years old we lived a different life, because they didn't cut down trees. We still made our money from farming—from bananas, coffee, cocoa, and sugar—but the river never dried up like it does now. Back then, the bare hills you see around us had forest on them. The river had plenty of water in it, so if we were hungry we could always catch crayfish. The woods had plenty of animals as well—deer and peccaries[1]—so people could hunt them too. And we didn't have to control the hunting, because there was plenty of everything."

---

1. Peccaries are the New World's answer to pigs. Also called javelinas, they are found from Arizona to northern Argentina. They evidently taste good too—a Google search for "javelina + recipe" yields 126,000 hits.

But in the 1960s, life in the four villages that make up the *comuna* of Loma Alta began to change. "A man called Juan Guales arrived and gave people the idea of making charcoal and of cutting down the trees for timber. He showed them how to build ovens for making charcoal. People saw it was easy to make money, and so they followed his example. People learned, and it expanded quickly. They used to get one and a half or two sucres [about US20¢] for a big bag of charcoal. Businessmen came from Guayaquil[2] and left advances, so people made more and more charcoal. But then we noticed the trees were getting scarcer, and people had to travel further and further to make charcoal. The hillsides nearby became bare, and soon the river started drying up."

Don Alejandro's account of events in Loma Alta is typical of what has happened, in living memory, throughout Ecuador west of the Andes— from the rolling rainforests along the border with Colombia, through the drier forests of the Colonche range, to the coastal deserts and scrublands that run south into Peru. At the end of World War II, almost three-quarters of this region was still under natural vegetation. But within 40 years that figure had collapsed to less than 5 percent, as unprecedented population growth, new roads and land reform programs opened up western Ecuador to timber extraction, charcoal harvesting, and farming.

Yet not quite all the forest has gone, and the day after meeting Don Alejandro I get together with fellow villagers Mauricio and Pascual Torres, Tanner, and a long-suffering mule for a three-day trek into one of the largest fragments still left in the Colonche Mountains. We leave behind the stilt-borne huts of the hamlet of El Suspiro[3] and walk through neatly edged fields toward the forested slopes of a small mountain called Cerro La Torre at the far end of the *comuna*'s land. The heat builds steadily; it's the tail-end of an 8-month dry season, but despite the clouds, there's not much chance of rain for a month at least. Gradually, the track steepens, the vegetation grows lusher, and I grow sweatier. Trees become more numerous and sport what seem to be strange outgrowths along their branches. Epiphytes. Plants growing on other plants: bright green ferns, shaggy clumps of moss, fleshy-leaved orchids with sprays of delicate white flowers, and most of all, the funnel-shaped forms of bromeliads, some with bright red flower spikes above them, all festooned with fringes of aerial roots hanging below.

We trudge on up the slope, the dry track giving way to mud as the

2. Ecuador's largest city, a 200-kilometer journey to the east.

3. Apparently named for the collective sigh (*suspiro*) the first settlers let out when they arrived there; whether it was a sigh of exasperation or of joy at the beauty of the place is not recorded.

lower reaches of the forest proper close in around us. Tropical forests are always subtler places than TV documentaries might have us believe—the birds harder to see, the spectacular and the abundant in reality usually shy and scarce. Yet this forest offers up some real treats to take my mind off the sticky and slippery climb. Jewel-like cicadas with turquoise-and-black bodies and delicately filigreed wings, the near-deafening calls of the males sounding like amphetamine-assisted tambourines. An iridescent green and flame-red dove-sized bird called a trogon. An anolis lizard, flat-pressed and precisely camouflaged on a moss-coated trunk. The gaudily extravagant scarlet and yellow zigzag candelabras of heliconia blooms. And the curious sight of walking leaves, which closer inspection reveals to be a dual carriageway of leafcutter ants. In one direction, workers struggle improbably to ferry their enormous loads back to the underground, leaf-nourished fungal garden that the colony cultivates for its food; the other way, their unencumbered sisters rush energetically back to harvest more foliage.

Our trail becomes steeper and the mud thickens, in places now calf-deep. I struggle, leave behind a Wellington, and fall ignominiously on my backside. Yet the mule plods steadily on, and with it our bedding and food. I notice the trees are becoming shorter, the undergrowth between them more tangled. Even though we're only about 600 meters above sea level we're entering what looks to me like cloud forest[4]—where the damp air protects the roots of epiphytes from drying out, and means many more plants can make a living with their feet out of the ground. Just about every branch carries a cloak of moss, plus a fern and a bromeliad or two for good measure.

At last we reach the crest of a rise and the hut where we'll spend the next two nights. We eat hungrily—me mesmerized by a black-and-purple tarantula so big that I hear the leaves it's rustling long before it crawls into view—and then set out to the very top of the ridge. Mauricio and Pascual are keen to show me hummingbirds (they've been helping research them here for over a decade) and to make sure we see plenty, they fill a set of feeder bottles with sugar water, hang them from the trees, and leave them for the birds to find.

When we come back the next morning we're not disappointed. The feeder bottles are partly red like the heliconias and bromeliads whose flowers have evolved to attract and be pollinated by the hummingbirds. Soon after daybreak they're already being fought over by the hummers. A me-

4. I'm used to seeing cloud forest growing only at much higher altitude—from 2,000 meters upward. But in small coastal ranges, which cool more quickly than bigger mountains and are more exposed to moist oceanic air, cloud forest can occur as low as 400 meters above sea level. This is known rather delightfully as the Massenerhebung effect.

tallic green and sapphire blur no bigger than yesterday's cicadas zips past my head: a violet-bellied hummingbird, Pascual tells me. It hovers at a feeder, wings ablur, body and elongated bill perfectly still, then darts on. A larger bird, more somberly dressed and with an even longer, sickle-shaped bill, perches on the rim: Baron's hermit, apparently. Looking around, I see each feeder now has two or three feathered jewels jostling for an energy-rich meal—a dazzling confusion of shimmering greens and metallic blues, their wings buzzing as they speed by, chasing away rivals. The names prove almost as exotic as their owners: the Andean emerald, the Amazilia, the green-crowned woodnymph, and the green-crowned brilliant. Their high-speed vitality and vigor are as captivating as their colors.

We see plenty more of Cerro La Torre's treasures over the next day and a half. Electric-blue morpho butterflies the size of my hand, tiny froglets hopping through the leaf litter, an excited family of red-tailed squirrels racing each other around some lianas, and an achingly beautiful king vulture soaring over the forest on huge snow-white wings. Pascual points out a shrub (a species of *Columnea*) with red-blotched leaf tips and explains how these work as long-distance signals to hummingbirds, who then find and pollinate the much smaller flowers tucked under the plant's big leaves.

In the afternoon, Mauricio tracks down the troop of mantled howler monkeys that woke us with their deafening dawn chorus but are now taking their siesta draped lazily across the sunny boughs of a canopy tree. Later he builds a makeshift tree platform where we wait, scarcely daring to breathe, for the user of the trail beneath us to pass by on its dusk-time foray. Almost an hour (and dozens of mosquito bites) later, a sleek russet-and-olive agouti—basically a lithe, elongated guinea pig—potters by, oblivious to our silent excitement. We're not as lucky staking out the paca (another outsize rodent, this one with stylish lines of white spots on its chocolate flanks) that Mauricio has somehow worked out also uses this trail. And a moon-bright search with a spotlight fails to turn up the campsite "kinky shoe" (an animal I later decipher as a tree-dwelling raccoon relative called a kinkajou). But we know they are there.

This forest is indeed a place of marvels. But it almost didn't survive at all. The changes that Don Alejandro witnessed and that have transformed so much of this corner of the world very nearly took the wonder-filled forest of Loma Alta too. Instead, it was saved. Yet this was not because of its immeasurable value for beautiful and bizarre animals and plants, but because of new ways of thinking about the forest's value for people. So what was this lifesaving benefit? There's a clue in the mud.

Diversity in Decline

Western Ecuador is extraordinarily diverse. Dramatic variation in rain-fall—from desertlike conditions near the coast to as much as 7 meters of rain each year near the Colombian border—have combined with a highly dissected landscape of mountains and valleys to make for a great mixture of natural habitats, with similar patches often isolated from one another. The result is a spectacular numbers of species, an exceedingly high proportion of which are endemic (that is, found nowhere else).

Botanists reckon the region probably has over 13,000 plant species (compare this with around 12,000 in the whole of Europe, an area 130 times bigger). Around one-fifth occur only in western Ecuador, and many of these are incredibly localized. Famously, when in 1978 world-class botanist Al Gentry[5] explored the 20-square-kilometer Centinela Ridge, he discovered 90 plant species known from nowhere else on earth. That's 30 times more endemic plants than are found in the whole of Scandinavia.

The diversity of the birds is similarly mind-boggling. Loma Alta has over 300 species, of which 79 are narrowly endemic (so-called restricted-range) species; by contrast the whole of the UK has just one—the Scottish crossbill. Almost 1,600 bird species are crammed into Ecuador (about twice as many as in the United States or the entire continent of Australia). The bird field guide for the country weighs 3 times as much as the equivalent book covering all of Europe, and should really come complete with a mule for lugging it around. And when it comes to learning the birds, while in my neck of the woods one or two words usually suffice to pin down a bird's identity (blackbird, dunnock, woodcock . . .), specifying Ecuador's mega-diverse avifauna often requires aristocratically triple- and even quadruple-barreled names (the crimson-rumped toucanet, the lachrymose mountain-tanager, the buff-throated foliage-gleaner, the scale-crested pygmy-tyrant . . .).

Yet in general, regions that have a lot of species (and especially endemics) also tend to become settled by lots of people. The foothills of the Himalayas, the east coast of Brazil, the islands of Southeast Asia, and the mountains of East Africa: all have very high biological diversity and large human populations. Scientists don't fully understand why this pattern arises, but

5. Al Gentry, fellow botanist Eduardo Aspiazu Estrada, and legendary ornithologist Ted Parker (a man who could apparently recognize 3,000 bird species from their calls alone) were all tragically killed in 1993 plane crash while on a reconnaissance flight for remaining forest fragments not far from Loma Alta.

it's generally bad news for nature. Because having many people around usually equates to plenty of pressure on natural systems, this association between species richness and human settlement leads in turn to disproportionate concentrations of threatened species—and western Ecuador fits the pattern exactly.

It was one of the first places to be settled anywhere in the Americas—by the Valdivian culture, which lasted from roughly 6,000 to 4,000 BP. Even now it's not hard around Loma Alta and elsewhere to pick up fragments of the Valdivians' characteristically chunky pottery. Other increasingly sophisticated cultures followed before the region's indigenous populations crashed in the wake of diseases brought by the Spanish. Recovery was then slow, but the post–World War II boom saw rural population densities soar and natural habitats disappear with alarming speed. In the time between Al Gentry visiting Centinela Ridge and his account of its extraordinary flora appearing in print, its forests were destroyed, and with them the only known populations of those 90 endemic plants. Nowadays, saving what's left of western Ecuador's biological wealth is seen as one of conservation's greatest challenges.

But as I've found everywhere on my travels, conservation is about people as well as biology. Social context really matters, and in those terms Loma Alta's forest might be expected to fare better than most. The land isn't owned by private individuals, so it could be argued that it's vulnerable to the so-called Tragedy of the Commons—with its resources being overharvested because everyone can take what they want, and if you don't overdo it then someone else will. However, as political scientist Elinor Ostrom has pointed out—in work that earned her the 2009 Nobel Prize in Economics— this view is overly simplistic. What matters is not who owns the land, but whether there are controls on access to its resources.[6] It turns out, Ostrom argues, that community-owned resources are often managed sustainably, provided the conditions are right. And on the face of it, several of those conditions are in place in the Colonche Mountains.

The community of mestizos—part Spanish, part Manta Indian—that arrived to farm here in the early 1900s had by 1937 been given legal tenure to the land, and so they knew it was theirs—they had a long-term stake in its future. The boundaries of the *comunas*, including Loma Alta, coincided with those of the watersheds, so that people's access to water, the most basic of natural resources, was in their own hands (rather than being depen-

6. So the Tragedy of the Commons should perhaps be more accurately (but less poetically) labeled the tragedy of open-access resources.

dent on the behavior of people living upstream). And *comunas* like Loma
Alta have a strong track record of democratic decision making, with a small
elected *cabildo* (council) making day-to-day decisions and deferring major
ones until there is agreement of the entire community. According to the
Ostrom school of thought, it's precisely these conditions—of strong property rights and good local governance—that should enable local people to
look after their jointly held resources sustainably. Once people are properly empowered, the argument goes, enlightened self-interest should do
the rest.

Nevertheless, as Don Alejandro recounted, when given the means to fell
their forest for charcoal and log it for timber, the people of Loma Alta did
exactly that. The *comuna* allocated villagers the right to cut particular trees.
It allowed them to clear areas for cultivating the palm-like plants whose
fibers are used to make panama hats.[7] And it let powerful ranchers from the
neighboring province of Manabí to come in and convert tracts of forest into
pasture for their cattle.

Conservationists did try to do something. A national NGO called Fun
dación Natura started trying to encourage local people to save the remaining forest. A U.S.-based organization helped the *comuna* in getting the national courts to recognize its right to protect its forest from outsiders. Yet
when U.S. conservation scientist Dusti Becker arrived in Loma Alta in the
mid-1990s to test Elinor Ostrom's ideas, trees were still being logged, there
was heavy hunting pressure, and the Manabí ranchers were continuing to
clear the forest—and all this in what was one of the highest-priority sites
for conservation on the planet.

Becker was perplexed. She obviously knew Ostrom's work well and knew
also that the people of Loma Alta had established clear tenure of their land
and were capable of making rules to limit its overexploitation. She knew
that conservationists had for many years told the community about the
value of their forest for biodiversity. So why were they apparently letting it
disappear? Maybe it really wasn't in people's best interests to keep the trees
standing. They could appreciate that woodcutters and fiber cultivators
needed to make a living, and while some people from far away told them
the area was biologically valuable, that wasn't by itself a good enough reason to constrain local livelihoods or take on powerful cattlemen. Or maybe

7. Panama hats are named after the country from where they were traditionally shipped, but
they're made in Ecuador, from the fibers of *Carludovica palmata* plants. These can be woven
so finely that high-quality panamas sell for thousands of dollars and can allegedly hold water
and be rolled up and pulled through a wedding ring (though not without spilling the water, of
course). This strikes me as an odd thing to do with very expensive headgear.

the forest was more important to Loma Alta's villagers than simply as a source of timber and land, but in ways that weren't yet apparent to the local community. The answer came when, of all things, Dusti was out on Cerro La Torre doing what she enjoys most—watching birds.

The Forest and the Fog

Dusti Becker is tall, blonde, humble, and cares deeply about the human as well as avian inhabitants of Loma Alta. She mockingly refers to herself as a giraffe—"especially among these people"—while for their part the villagers are evidently very fond of this researcher who returns for several months each year and works so hard for their forest. When I catch up with her, she's making her Christmas visit to El Suspiro, and in between handing out presents to innumerable godchildren and listening to their news, she tells me about her "eureka" moment on the mountain.

"There were about 10 of us on a birding expedition. It was the dry season, but we were high up in the forest, and it was foggy. It was so wet all around us it felt like it was raining, but it wasn't. Then we walked into the pasture [cleared by the Manabí ranchers] and it wasn't wet at all." The soil was different too. "In the pasture the soil was dry, almost like sand, but in the forest it was mud—even though it hadn't rained."

Suddenly she realized the importance of the vegetation. In the forest she and her fellow birders were getting wet because the aerial tangles of leaves, stems and roots were intercepting the air's moisture, which was then dripping onto the forest floor, the expedition personnel, and their notebooks. This wasn't exactly remarkable—that's how cloud forests work—but two other things struck Dusti. First, this wasn't happening in the pasture: where the forest had been removed and replaced with shorter and structurally simpler vegetation, the moisture passed through uninterrupted. Even more significant, the forest trees and their multilayered epiphytes were capturing water and dripping it down the necks of her crew not in the rainy season—like in most cloud forests—but in the dry season, when fog (known locally as *garúa*) rolls in off the Pacific, sometimes for weeks on end.

By harvesting water when there was little prospect of rain for many months, maybe the remaining forest of Loma Alta was performing a very important service for all the villagers. Maybe it was keeping the mountain soils wet and the streams that drain them flowing, long into the dry season. And maybe that was why, as Don Alejandro had observed, the river started to dry out after people began cutting down the trees for charcoal. "We all got so excited by the idea we started going crazy," Dusti recalls. But she also knew forest hydrology was a lot more complicated than people sometimes think, and so decided to test her idea with some data.

There's a widespread belief—among the public but also among some policy-makers and even conservationists—that having trees around increases water flow. Cut trees and streams dry up; replant them and the rivers run once more. The problem is, in a general sense, that's almost completely wrong. As good gardeners know, and as South African farmers discovered when all those antipodean eucalypts and wattles escaped into the fynbos, trees drink a great deal more than other plants. When the vegetation consists of grass or shrubs rather than trees, the total volume of water lost from leaves via so-called evapotranspiration plummets. So in most instances, the total amount of water left to flow down the streams that drain a hillside increases if it's deforested.

There are some serious (albeit more subtle) hydrological downsides to losing trees. By becoming more exposed to the weather, fragile forest soils are more likely to be eroded, which lowers downstream water quality and also reduces the ground's capacity to soak up moisture and release it slowly. As a result deforestation can sometimes increase wet-season flooding, and decrease vital dry-season flow. But in terms of the total amount of runoff over an entire year, having trees around is generally agreed to reduce water availability. The one, still controversial exception? Cloud forests, which because of their complex three-dimensional architecture may actually strip out more water from the air than they give back. Keeping cloud forest may help maintain runoff.

To find out whether Loma Alta's garúa-fed forest helped or hindered water flow, the next dry season Dusti trained a team of foreign volunteers plus her two versatile assistants, Pascual and Mauricio, to measure the total amount of moisture falling directly from the sky (usually as mist) and dripping off the vegetation in different parts of Cerro La Torre. In order to quantify the effects of forest clearance, they placed cylinders in the pasture and in nearby forest close to the top of the mountain, as well as in two panama-hat-plant clearings and two neighboring forest patches lower down. The results were striking.

Dusti explains, "Some days the cylinders in the clearings had so little water we got it out using medicine droppers. But the forest cylinders always had plenty." Over the course of 14 typically foggy days Dusti's team discovered that the cylinders in the forest collected three times as much as cylinders placed the same height up the mountain but in panama-hat plantations, and 6 times as much as the same-altitude cylinders in the pasture.

Because all the cylinders were exposed to roughly the same amount of fog falling straight from the sky, this huge variation in the water they captured must have been due to differences in fog dripping off plants. In the forest, water was dripping into the cylinders from the vegetation overhead, whereas in the clearings, the fog-bound moisture passed through unintercepted. So her insight from the year before was correct: losing the forest was indeed reducing dry-season fog capture. But was this effect big enough to matter to the people of Loma Alta?

To work this out, Dusti scaled up the figures from the cylinders, based on the number of foggy days in an average dry season and the observation that ranchers had by now converted around 200 hectares of Cerro La Torre's forest to pasture. Even assuming conservatively that only 10 percent of foregone fog capture would have become available for downstream use by the villagers, this calculation suggested that the loss of water to the *comuna* because of clearance for pasture was around 38 million liters every dry season—or (expressed in the standard units for comparing hard-to-imagine volumes) enough to fill 15 Olympic-sized swimming pools. But because dry season water is so scarce in Loma Alta, it's bought and sold, so Dusti was able to go one stage further and estimate the annual value of the lost water: around US$130,000. In a community with only 200 households, this was equivalent to around half of the average annual family income. Clearly, losing the fog interception service provided by the forest was very important indeed.

Dusti and her colleagues quickly showed their findings to the elected *cabildo*, who then organized for the group to make a video presentation to an *asamblea* (or decision-making gathering) of all four villages. The film showed the team collecting fog data and explained that recent declines in village water supply could be the result of forest clearance. Wilson Tomalá, president of the *comuna* at the time, remembers the discussion. "The science showed us how much water was lost from deforestation. The answer was very alarming. We were running out of water, so we realized we needed to change."

The villagers took a vote on establishing a forest reserve where further clearance would be banned and logging and hunting prohibited. Eighty-three percent were in favor. "But people who made a living out of the forest—the woodcutters, the charcoal-makers and the hunters—were against the idea," the former president explains. So the *cabildo* deferred a decision in order to give more time to hear the concerns of the most directly affected families. The *cabildo* offered them first preference for new jobs, as forest guards. The *asamblea* met again a week later, and then again, until

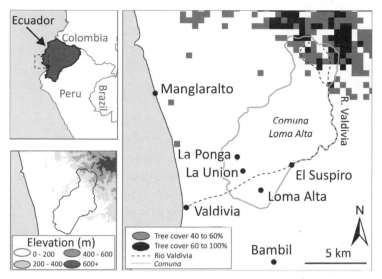

Map 6A. The people of Loma Alta in western Ecuador have acted to safeguard their high-altitude fog forest and hence their way of life. Their neighbors in Bambil have not. Tree cover data from Hansen et al. 2006; *comuna* information from Becker 1999; and elevation data from NASA (STRM 90m data set provided by CGIAR at http://srtm.csi.cgiar.org). Cartography by Ruth Swetnam.

consensus was reached. Finally, after four meetings, the *comuna* unanimously declared 1,000 hectares of its land as the Reserva Ecológica de Loma Alta. What conventional conservation exhortations had failed to achieve in more than a decade, the focus on fog had delivered in less than 3 months.

Pride and Progress

It's now 10 years on in El Suspiro, and just before the equatorial dawn. I'm meeting up with a bunch of bleary-eyed but eager village youngsters, half of whom weren't born when the reserve was declared, for the annual Loma Alta Christmas Bird Count. These exercises began in the States over a century ago as an alternative to competitive yuletide shooting parties. They've grown to the point where they now provide useful data on trends in bird numbers across almost 2,000 places in the Americas, but today's count is 1 of only 3 in the whole of Ecuador.

The groups disperse, and ours heads up the road toward the hills, the knot of children chattering excitedly and with Dusti leading the way like the Pied Piper. Our job is to spend the morning recording all the bird species in the hinterland between the village and the start of the forest; other, more experienced groups are heading for the forest proper. We peer through shared binoculars as Dusti explains the features that make this bird a Baird's (and not a boat-billed) flycatcher, that one a southern yellow grosbeak. She points out bird calls and the children listen, trying hard to learn. "*Sa-co las lla-ves*,"[8] they repeat in singsong voices, mimicking the fasciated wrens in a thicket.

The children's curiosity spreads to other creatures besides birds. One boy giggles as he spots a wriggling cluster of tadpoles in a stream. Another finds a young tapiti (or Brazilian rabbit) hiding in a thicket. Dusti shows us a meter-long iguana straddling a branch then trumps it with an utterly dazzling emerald-and-turquoise jay-sized bird perched confidingly on the edge of nearby bush. Its long tail, ending in two spatula-shaped racquets, hangs below. As the bird starts to call, it swings its tails backward and forward in perfect time. Tick-tock, tick-tock. "Blue-crowned motmot," Dusti informs me. "But the kids have a much better name: *relojero*—the clock-bird." We spot loads more on our survey. Tiny parrotlets, a mockingbird, vultures, a pygmy owl. Altogether the children record over 70 species in a few hours. Much more importantly, they have a lot of fun.

Over the decade since its creation, the reserve has flourished. The villagers have chosen to expand it twice. It now covers 3,000 hectares, which is more than 40 percent of the *comuna*. Former president Tomalá says, "The more we understood why we were protecting the reserve, the more we realized it needed to be much bigger." It's a commitment that shows an exceptional level of self-restraint. While the rest of the (often much richer) world struggles to meet international targets to set aside 15 percent or even 10 percent of their land area for nature, Loma Alta's inhabitants have understood exactly why conservation matters for their livelihoods, and their children's, and have gone three times as far.

Enforcing all the rules has taken a while longer. Families with hat-fiber plantations in the reserve have been allowed to continue cultivating them, but new clearance for pasture for hat plants was quickly stopped. Bringing logging and hunting to an end has proved trickier, with successive *cabildos* reluctant to impose the reserve's regulations. But in recent years these restrictions, too, have been enforced. The main woodcutter has bought a truck and now makes his living as a taxi driver–cum–hat-fiber transporter. The current *cabildo* president tells me proudly that not a single tree has been cut in the reserve for almost a year.

External encouragement has been important in maintaining local back-

8. This translates roughly as "I take out the keys. . . ."

ing for the reserve. Early on, PAN (People Allied for Nature, one of the organizations that supported Dusti's fog work) installed a new water pump in El Suspiro. They provided legal help to end the dispute with the Manabí ranchers, funded assistant teachers in the local elementary schools, and now pay for the *comuna*'s children to attend high school in Valdivia. When the woodcutter swapped his chainsaw for a taxi, PAN helped his daughter get a place there.

PAN also brought in craftsmen to teach villagers how to carve the hard, white insides of "vegetable ivory," or *tagua* nuts—collected from palms[9] growing wild in the forest—into earrings and other souvenirs for visitors. Dusti's organization, Life Net, brings in a few dozen people each year to monitor birds and has a strict policy of employing only local people as assistants, guides, and cooks. There's a trickle of other visitors, nationals as well as foreigners, who come for the wildlife and the archaeology. Between them, *tagua* carving and tourism have replaced woodcutting and hunting in the local economy.

After all of the Christmas bird counters have returned, we crowd into the village hall—which already seems to have half the village in it—and Dusti goes through the results. There are cheers when each group's total is declared, when names of the scarce or rarely seen are called out, and especially when the combined total is announced. Two hundred and one species—a new record—in little over half a day. Afterward, Fundación Natura puts on a puppet show involving a giant tree, the good fairy of conservation, and some not-so-good would-be woodcutters. From what I can work out, the story's about the forest and why keeping it is a good thing, and it looks like it's pretty funny, at least if you're under 10 and speak decent Spanish.

Environmental education and a culture of pride in the natural heritage of their lands are at the core of the current *cabildo*'s vision for Loma Alta. Its dynamic young secretary explains to me that he wants to make sure the next generation understands the value of the *comuna*'s forest—that if it's properly managed it can provide jobs as well as water, into the long-term. Strong protection, he believes, can secure a sustainable future for Loma Alta. It's a popular message: the *cabildo* has just been reelected for another year.

But while pride and jobs are clearly key parts of the package, it was thinking about water that altered the fate of the forest and, as a conservation scientist with only a rudimentary grasp of forest-water relations, I still have a

9. In honor of their nuts, this genus of palms is called *Phytelephas* (or, elephant plant).

nagging question to ask Dusti. A couple of the experts I've asked have cautioned about the difficulties of measuring how much fog really is trapped by a forest. Some of what gets recorded as fog might actually be slanting rain, and some of it gets evaporated by plants before it soaks into the soil (so is never of any use to people). The only reliable way to measure the effect of forest cover on downstream water availability, they argue, is to install flow-gauges in several streams that drain still-intact forests and several others that drain recently cleared areas, and compare the results.

Dusti has careful counterarguments to each point. In the *garúa* season there's virtually no rain, slanting or otherwise, to confuse things; the fog means that humidity is very high, so evapotranspiration is low; her estimate that only a miserly 10 percent of the intercepted fog makes it downstream is pretty conservative; and collecting simultaneous flow data for multiple streams would have been great, but taken far more money and time than were available. But forget the niceties of understanding exactly what's going on in the fog forest. "If you really want to see what would have happened if the people of Loma Alta had cut down the rest of their trees" she says, "it's easy. Just go to the *comuna* next door."

To Have and Have Not

A 15-minute motorbike ride from the neatly tended fields of El Suspiro and I feel like I've landed on the moon. We travel along what was once a riverbed but is now a deeply eroded gully, bone-dry and thick with dust. We round a bend and stop to take in the scene. In all directions the ground is ash-gray, parched, near-lifeless. Dead shrubs, gnawed down to bare wood by goats, are scattered skeleton-like across the hills. The only green patches I can see are clumps of cacti. The skull of a long-dead cow lies by the roadside, an echo, maybe, of a better time.

We're just one ridge and a few kilometers from Loma Alta, but a world away from the lively, forward-looking place where I've spent the last few days. As we approach the village of Bambil Callao we pull over to let a lumbering pickup overtake us. It's the water truck, and it comes here five times a day, every day. The water originates in Loma Alta, and without it, this village would die.

We track down the president of the Bambil *comuna* and sit together on white plastic chairs outside a bar. The village reminds me of a set for a spaghetti Western—dust-swept, deathly quiet, and with a couple of turkey vultures wheeling lazily overhead. I ask about the water truck. "We haven't had any water of our own for 30 years. Instead, people pay the water carrier 50¢ to fill a drum in their house."

He points to the bare hills beyond and tells me that there used to be trees, "but we cut them all down for timber or charcoal. We even had a fog forest once. And we used to farm, when the river was still running." But unlike at Loma Alta, the cutting here never stopped, and the water ran out. "There's no farming at all now. Instead, everyone has to work in the tuna-canning factory two hours away in Posorja. Thirty years ago things were much better because people could make their living here—they didn't need to go away." I look around and realize I can see no people of working age whatsoever.

The president shifts in his chair. I get the feeling he doesn't want to talk any longer, but as we get up to go, he looks over to the ridge we've just come across. "It's good that they're now protecting their forest in Loma Alta. It's good. Here, we have no forest left at all. We have no chance."

Back in El Suspiro I ask if there are other places in the same mess as Bambil. La Balsa, and another dried-up hamlet optimistically called Barcelona: these too have lost pretty much all of their forest, and depend on the water truck from Loma Alta. "They feel bad because they went too far," Mauricio tells me. "They can't farm because the river is dry. They can't do anything. They wish they could go back, but they can't." But other *comunas* nearby have seen what's happened, and what Loma Alta has managed, and they're trying to change things before it's too late. In Dos Mangas, in Olón, in Febres Cordero: "They're now starting to protect the forests they've got left, for the same reasons." The leaders of Loma Alta are doing what they can to help.

Such recognition of the hidden values of ecosystems is catching on elsewhere too. It's obviously an important part of the motivation behind the Dutch National Ecological Network. It's underpinning radical approaches to managing river catchments in New York State and in parts of the UK. Worldwide, thinking about ecosystem services has become one of the most pervasive strands in contemporary conservation. The key, its proponents argue, lies in seeing a patch of nature not just in terms of the intrinsic worth of its birds or trees—important though that is—but in terms of people, and how they may stand to gain from conserving the system. Where those benefits are great enough to outweigh the advantages of turning trees into charcoal or draining a wetland to grow food, then quantifying nature's services can help build powerful new arguments and win over influential champions for conservation.

In Loma Alta, the step from characterizing the main benefit (in this case, fog capture by the cloud forest) to influencing the behavior of people on the ground was relatively straightforward for two reasons. First, most people here feel they have enough land to meet their needs, and so the cost to them of not converting the forest—what's termed the opportunity cost of conservation—is relatively modest. Second, because of the alignment of the *comuna* system along watershed boundaries, the water captured by the forest flows to the same community who controls its fate: the main beneficiaries from its conservation are thus also the main decision-makers. Even then, some outside support—in this case from people far away who get satisfaction from knowing the forest is still intact (and so contribute to NGOs like Fundación Natura and Life Net)—has been important in tipping the balance of costs and benefits in favor of keeping the forest.

But elsewhere, things are often different. The gain from converting a habitat (or put another way, the opportunity cost of keeping it) may be greater; and many of the main benefits of its conservation may be what economists call externalities, meaning they are enjoyed only by people elsewhere, who don't own or manage the habitat. For example, it may be locally profitable to clear a mangrove swamp to make way for a shrimp farm, and the resulting loss of ecosystem services (such as the nurturing of young fish that then disperse to an offshore fishery) may not be felt by the owners of the area. Likewise, commercial ranchers or soybean farmers can sometimes make a lot of money from turning tropical forest into farmland, while the people who lose out, because the release of carbon previously stored in trees exacerbates climate change, are scattered all around the world.

In these more common circumstances conservation depends on whether the loss of such external benefits becomes a material consideration for people who own or control tracts of habitat. The conventional conservation solution is regulation—laws restricting habitat conversion. But recognizing the value of ecosystem services offers an alternative approach, based on carrots instead of sticks. Once the main beneficiaries of conservation have been identified, they can offer incentives to local owners or decision-makers to persuade them to continue to provide the services they want (nurturing young fish, absorbing carbon dioxide . . .). In economic jargon, these incentives serve to internalize the externalities of conservation: they make them apparent to the people determining what happens on the ground.

In most cases—not in Loma Alta but in most other examples—the incentives consist of money. And one country has gone further than any other in turning this idea of paying for ecosystem services into a reality.

Repaying Natural Capital

Every time motorists in Costa Rica fill up their cars with fuel they inadvertently help save their country's spectacular forests. The same is true for many Costa Ricans when they pay their water or electricity bill, buy a beer, or catch a plane. A compulsory tax on fossil fuels,[10] and optional levies agreed by municipal water suppliers, hydroelectric companies, and bottling plants, as well as both domestic airlines, all go toward financing a program of payments for ecosystem services (called Pagos por Servicios Ambientales, or PSA) first introduced in 1997 and recognized as the most ambitious scheme of its kind anywhere in the world.

Costa Rica is a pioneering place. Since 1948 it has had no standing army. Instead (and unlike most of its Central American neighbors), it's had decades of democratic government, steady economic growth, and serious investment in education. It now has among the highest literacy rates in the Western Hemisphere, less corruption than the Czech Republic or Greece, and greater press freedom than France. According to the New Economics Foundation its citizens are the happiest on the planet.[11] Yet Costa Rica's postwar development came at a high environmental price, with tax breaks and other government policies to encourage agricultural expansion leading to one of the highest deforestation rates in the world, and the loss of almost two-thirds of the nation's remaining forests in just 30 years.

By the late 1970s the government decided to act, putting subsidy schemes that promoted forest conversion into reverse and instead paying millions of dollars to protect remaining forests and replant cleared areas. The results were impressive—Costa Rica became the first developing country to slow deforestation to the point that forest regeneration began to outstrip clearance. But under pressure to cut public spending, this level of direct government support became financially and politically unsustainable. Forest

10. Currently 3.5 percent of market value.

11. Corruption data are from the annual Corruption Perceptions Index of Transparency International (http://www.transparency.org/about_us), which ranks New Zealand as the least corrupt country and Somalia as the most. The press freedom scores are from Reporters Without Borders (http://www.rsf.org/) in their Worldwide Press Freedom Index. The happiness rankings come from the Happy Planet Index (http://www.happyplanetindex.org/), which calculates "sustainable well-being" from a country's average life expectancy, the reported happiness of its citizens in a globally standardized poll, and their average ecological footprint. Costa Rica ranks 1st out of 143 nations surveyed, the Netherlands 43rd, and the United States 114th.

conservation had to be funded another way—preferably by the people it benefited most.

The solution was the groundbreaking 1996 Forestry Law, which combined the hefty stick of making all deforestation illegal with the carrot of the new PSA scheme. This formally recognized that Costa Rica's forests provide four key ecosystem services—storing carbon, protecting water supplies, conserving biodiversity, and providing scenic beauty—for which their owners should be rewarded. When I interview him, Carlos Manuel Rodríguez—one of the chief architects of the policy and former Minister for Environment and Energy—remembers Congress asking him whether a deforestation ban could really be enforced. "We said, if at the same time the beneficiaries of conservation pay for carbon fixation and other ecosystem services, then landowners would see forest conservation as something that could compete economically as an alternative to farming." He was right.

Money to pay for the PSA program came from different beneficiaries of the services provided: fossil fuel users in Costa Rica and Norway (who gain from the carbon soaked up by the forests),[12] irrigation, bottling and hydroelectric companies (reliant on river water from forested catchments), and the Global Environment Facility (a major international fund set up to finance biodiversity conservation). "At the time we didn't know how much the different services were worth or how they varied in different types of forest," explains Señor Rodríguez. So rather than trying to work out the value of each service provided by any given piece of land, payments were made for all services combined, at a flat rate right across the country.

Landowners were initially paid around US$40 annually for each hectare of remaining forest they undertook to conserve,[13] and up to US$550 over 5 years for reforesting a hectare. Even at this fairly modest level the scheme was soon oversubscribed, with applications to join up far outstripping available funding. By the end of 2005, around a quarter of a million hectares—roughly one-tenth of the country's forest area—was under PSA contract.

Most people I've spoken to think the scheme has been a success, but pinning down exactly what it's achieved turns out to be a bit tricky. Since PSA

12. These are the two most populous countries to have so far made formal pledges to become carbon neutral, both by 2030. Of the other countries that have signed up, Iceland plans to do so through geothermal and hydroelectric power, the Maldives through wind and solar energy, and the Vatican by planting a lot of trees in Hungary.

13. This was roughly comparable to the return from cattle production, which at the time ranged from US$8–125 per hectare per year. PSA payments have since increased to around US$70 per hectare per year.

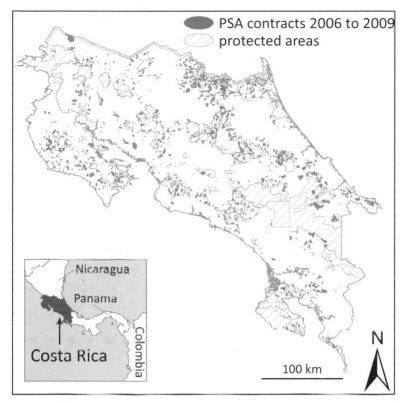

Map 6B. Participation in Costa Rica's ground-breaking PSA program supports the protection of privately owned forest across the country. The map shows all contracts signed between 2006 and 2009; many are in areas with limited coverage by formal protected areas. Data provided by Carlos Manuel Rodríguez and Gilmar Navarrete Chacón. Cartography by Ruth Swetnam.

was set up, the conversion of remaining natural forests has pretty much ceased altogether, while reforestation and natural regrowth have accelerated to the point where total forest cover in 2005 was estimated by some to have more than doubled since its mid-80s nadir. It's tempting to ascribe this remarkable turnaround directly to the fact that landowners are for the first time now being paid to keep their forests standing—and several surveys show, as you'd hope, that farms signed up to the scheme do indeed have more intact and more recovering forest than non-PSA properties.

But it's not quite that simple. The recovery of Costa Rica's forests began well before PSA started, may not be quite as marked as some people claim, and is doubtless partly down to the outlawing of deforestation and the reversal of subsidies for forest clearance. It's also likely that pro-conservation landowners have been more willing than other farmers to join the program (and would have looked after their forests better even in the absence of payments), so the difference seen in forest cover on PSA versus non-PSA land isn't necessarily entirely because payments altered farmer behavior. One recent, clever analysis that compared PSA farms against carefully matched land with similar owners who didn't enter the scheme has found evidence that the program has slowed forest loss and speeded its recovery. But the most sensible interpretation, it strikes me, is the view that PSA has been a vital sweetener in an overall package of sticks and carrots that together have changed the fate of the country's forests. As Carlos Manuel Rodríguez argues, "The laws which banned land-use change wouldn't have been enough—but PSA made them politically acceptable."

Looking to the future, it's clear that Costa Rica's pioneering program faces important challenges. It needs a broader funding base within the country, with less reliance on government support and international loans and more money flowing from Costa Rican beneficiaries of the services it's striving to protect. Rodríguez sees a solution in ongoing efforts to introduce a levy on all water users, which he thinks will bring in as much revenue as the fossil fuel tax within a few years. There's also a widely acknowledged need for greater targeting of payments. The flat-rate system is inevitably more attractive in areas where the opportunity costs of conservation are low—but that's where forests are under least threat. On the other hand, differentiated payments with higher rates in areas at higher risk would have more appeal to farmers who would otherwise be tempted to clear their land.

More sustainable funding and better targeting should increase the impact of the PSA scheme on the ground, but it's already had a long-term impact somewhere equally important: in the minds of Costa Rican people. By recognizing the importance of ecosystem services and by enabling those who benefit from them to reward those who provide them, the PSA program has changed the paradigm of forest management. It's brought conservation into the mainstream, and shown that it can be compatible with good business practice and continued economic growth. As ex-minister Rodríguez puts it, "We proved a developing country can succeed using conservation as an economic engine. We showed that an acre of forest is worth more than a cow."

Costa Rica's example of those who gain from ecosystem services paying other people to manage (or leave well alone) areas of nature so that they continue to generate those services has caught on around the world. There are service payment schemes springing up in Asia, Africa, Europe, and North America, as well as across Latin America, from Chile to Mexico. But my favorite example, for its lateral thinking and wit and for achieving the near-impossible task of successfully tackling three problems at once, comes from another and far-distant corner of the world: the near-uninhabited, sweltering wilderness of Australia's Northern Territory.

Fighting Fire with Fire

On the morning I get hold of Peter Cooke on the phone, Britain has just endured its coldest night in 30 years. It's gotten cooler where Peter is too—"It's only 34 degrees centigrade, now that the rains have come," he tells me from his safari tent at the remote outstation of Kabulwarnamyo in western Arnhem Land. But while people in Britain have become briefly obsessed with the perils of a few days of ice, Peter is a man concerned with a very different, much more persistent challenge: fire.

"There's a 50,000 year tradition of managing the Australian landscape using fire." Peter explains how, in what's been labeled "fire-stick farming," aboriginal people roamed large areas of savannah lighting frequent fires to drive animals like kangaroos into the path of hunters, expose recent tracks (and hence, whereabouts) of burrowing animals, and attract game to the fresh grass that sprouts soon after burning. Because the fires were often lit early in the dry season before large amounts of dry fuel had built up, they tended to be small, but they also acted as firebreaks, limiting the spread of later burns. The consequence was a mosaic of different-aged patches of recovering vegetation: ideal for hunting and a great though inadvertent—way of keeping a diverse, species-rich landscape as well.

With the advent of Europeans, all this began to change. Late nineteenth-century colonization of the Top End (the vast squarish lump of land jutting out of Australia's northern coast) saw aboriginal populations collapse through exposure to unfamiliar diseases such as smallpox and violent conflicts with new settlers. It didn't stop there. "Drawn by the lure of jobs and tobacco, most remaining aboriginal people moved off 'country'" [their ancestral lands] "to work in European-run mines, mission stations and buffalo camps."[14] By the mid-twentieth century the entire western Arnhem Land plateau—a central chunk of the Top End the same size as Belgium—was almost empty.

14. These were camps for hunting feral water buffalos, which were introduced from Asia as a food source for early European settlers and by the twentieth century were abundant on the floodplains northwest of the Arnhem Land plateau. They're now recognized as a serious threat to native biodiversity.

Depopulation had profound consequences. Without deliberate ignition by people, fuel loads built up. Just as Smokey Bear's campaign led unintentionally to great conflagrations in the United States, the cessation of frequent burning in the Northern Territory meant that when fires did light they were far hotter, larger, and longer lasting. Mid- and late-dry-season fires often burned for months, raging over tens of thousands of square kilometers until the rains finally put them out. By the time people started worrying about these things, savanna fires had become responsible for over 40 percent of the Northern Territory's emissions of greenhouse gases.

The exodus had other impacts too. According to Peter Cooke, "A lot of country began to suffer from the absence of traditional burning." The region's species-rich, monsoon-maintained jungles, which stand out like islands in a sea of savannah, shrank back in the face of hotter, more extensive burns. Many of the endemic plants found on the plateau's unusual sandstone heaths declined. Like South Africa's fynbos plants, these depend on patchy fires that allow them to germinate and reseed between burns but have been unable to cope with the new, fiercer fire regime. And there have been human costs too: in the oldest surviving culture on earth—one whose intimate attachment to the land is reflected in hundreds of names for vegetation types and dozens of words for different rock formations, winds, and fires—aboriginal elders have become increasingly concerned about their people losing physical contact with their country.

In the late 1990s, Cooke (who'd long worked on ways of helping people "return to country"), fire ecologist Jeremy Russell-Smith, and renowned aboriginal leader Wamud Namok[15] began to work out a solution that, remarkably, promised to help local people, nature and the global climate. Using federal government funding they discovered that early, patchy burns released far fewer greenhouse gases than the hotter, more extensive blazes that had replaced them. Suddenly there was the prospect of reinstating traditional land management while simultaneously tackling climate change.

Peter elaborates on how they developed a "two tool-kit approach"—combining aboriginal skills and knowledge with Western science: the oldest culture working alongside space-ace technology. "On the aboriginal side we built capacity to deliver old-style fire management. On the science side we developed credible ways of measuring the consequences for greenhouse gas emissions. Our strategy was to emulate what aboriginal people did in

15. Wamud Namok died in 2009, and because his people consider it inappropriate to say the names of the recently dead, I refer to him here by his "sorry name." It's simply the name of the subsection and clan to which he belonged.

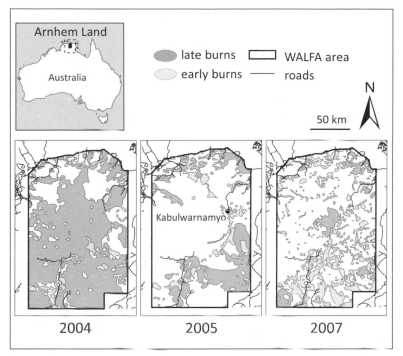

Map 6C. Since WALFA's inception in 2005, aboriginal fire managers have greatly reduced the extent of intense, late-dry-season fires by lighting early-season firebreaks. Map data provided by Stephen Sutton, Bushfires NT. Cartography by Ruth Swetnam.

the early dry season—burning as they traveled and incidentally putting in firebreaks. It's very rugged and inaccessible country, and there are far fewer people than before, so we used helicopter-borne incendiary devices to start fires and satellites to show us where to burn in order to separate different compartments with firebreaks. Meanwhile, the scientists were working with aboriginal rangers on the fuel loads of different vegetation types, devising a scale to estimate the amount of greenhouse gases these would release, depending on when they were burned." With proof of the principle of strategic fire management established, the next step was to secure long-term funding.

The answer came when U.S. energy giant ConocoPhillips applied to build a major liquefied natural gas plant on a piece of jungle bordering Darwin Harbour. The Northern Territory government told them they needed to find a way of offsetting some of the plant's greenhouse gas emissions, and do something to benefit those fire-stressed jungles. In response, ConocoPhillips agreed to pay the West Arnhem Land Fire Abatement project AU$1 million annually to unroll strategic fire management across 28,000 square kilometers of western Arnhem Land. The WALFA team—led by Peter, Jeremy and Wamud Namok—reckoned that by doing this they could reduce greenhouse gas emissions by around 100,000 tonnes of carbon dioxide equivalent[16] each year.

All the evidence is that WALFA is working. Satellite images show that through a combination of aboriginal people lighting small, patchy fires on long-distance bushwalks and helicopters burning strategic firebreaks in more remote areas, only 10 percent of fires are now of the really damaging, dry-season variety—down from one-third in just 4 years. Best estimates are that over the same period the project has managed to lower emissions from the plateau by almost 500,000 tonnes of carbon dioxide equivalent—some 20 percent ahead of target, without even counting any overall loss in emissions of carbon dioxide itself.

There are positives for aboriginal people as well. "At the start of the new millennium the government turned away from helping people return to country," Cooke complains. "Instead, they encouraged everyone to go to town and learn to be a hairdresser or start a coffee shop. It was pretty depressing. People who hadn't already gone back no longer had the support to do so. But this project is now enabling some people to return. It's created jobs in places where there weren't any before." It's also catalyzed the transfer of traditional knowledge from elders to generations who've lived entirely off-country and built new and positive relationships between aboriginal and European communities. Prospects for nature are on the up too. Peter talks about hard-hit stands of cypress pine no longer declining. "Other areas that were suffering badly from fire—especially jungles—are also doing a lot better. And we know that in sandstone heath the plants are getting much more opportunity to mature and set seed."

WALFA is proving such a success that other groups are now raising funds to establish indigenous fire abatement projects elsewhere in northern Australia—optimistic that these too can simultaneously achieve major emissions savings, help aboriginal communities, and conserve biodiversity. Proponents are also suggesting there may be scope for similar projects in African savannahs. Meanwhile, Peter Cooke is realistic about what

16. Bushfires release other, more potent greenhouse gases as well as carbon dioxide—in particular, nitrous oxide and methane—whose concentrations can be expressed in carbon dioxide equivalents (that is, the amount of carbon dioxide needed to have the same greenhouse effect). Because strategic fire management mainly delays (rather than eliminates) carbon dioxide release, the WALFA target deals only with nitrous oxide and methane. There is evidence of a long-term reduction in carbon dioxide emissions too, but this is an added bonus not included in the 100,000-tonne figure.

WALFA itself can achieve. "A million bucks isn't a lot of dough any more. What we're doing won't remove aboriginal poverty, and there are a lot of other land management issues that still need to be dealt with. But there are very few economic opportunities going to come up in this part of the world, and no others that are so kind to country. What we're doing is a lot better than nothing."

Today Arnhem Land, Tomorrow the World?

The incorporation of ecosystem services into conservation thinking and particularly the idea of paying people to look after land (or water) in ways that secure the provision of services is spreading fast. Not surprisingly, it's a change in emphasis that raises serious concerns among conservationists. One commonly held fear is the danger of commoditizing biodiversity—that putting a price on nature undermines other, moral arguments and makes the case for conservation vulnerable to changing technologies and human needs. For instance, if scientists invent a cheap technical fix for extracting large quantities of carbon dioxide from the atmosphere so that the price people are prepared to pay for natural carbon storage plummets, what then happens to the case for conserving tropical forests, which may now be worth less left standing than converted into soybean plantations? Those who live by the sword of economic rationality die by it too.

Personally, I don't find this objection overwhelming. The economic case for conservation should only ever be one of several arguments, and society could (and sometimes does) choose to conserve species and places (or rare books or significant works of art) even when it doesn't make economic sense to do so. But a fairly long list of more subtle concerns undoubtedly limits the usefulness of an ecosystem service focus in general, and of payments for ecosystem services in particular.

There are still many parts of the world where people are not yet plugged into the market economy, and are simply not well placed to deal with significant flows of cash from the outside world. On the other hand, the scope for a better understanding of ecosystem services to change the behavior of key decision-makers without money moving around may be pretty limited: it depends, as at Loma Alta, on substantial benefits accumulating locally.

We also have a lot to learn about economic values: for many potentially important services, like nutrient cycling and protection from flooding, economists and scientists are only just beginning to work out what they're worth and how they may be affected by different land management decisions. There are still no markets for most services, so setting up payment schemes usually requires legislation to encourage beneficiaries to pay. Maximizing one service—such as carbon storage—may well reduce an area's capacity to deliver other services (all those trees are likely to use up a lot of water . . .), so trade-offs and compromises are likely to be common-place. There's a real need to monitor the provision of services over time (so people know they're getting what they're paying for), and—as in Costa Rica—for better targeting of payments toward places where they're most likely to make a difference. And last, given all these difficulties, most pay-ments will usually be quite modest, so by themselves they're only likely to change destructive practices in places where the opportunity costs of con-servation are low as well.

All that said, my own view is that despite these caveats, thinking about ecosystem services can help highlight the hidden benefits we get from na-ture and bring in major new sources of conservation funding. In a world where conservation is limited by a shortage of cash and where we usually ignore what we can't see until it's too late, that's good news. And focusing on ecosystem services has the potential to do one other vital thing: to bring conservation into the mainstream, and so draw in actors from the private sector, as well as government and NGOs. In the next chapter, when I man-age to get to Australia in person rather just call it up on the telephone, I discover just how far commercial operations might go in contributing to conservation on the ground.

Figure 14. Epiphytic plants like these bromeliads and mosses play a key role in intercepting dry-season fog. Photo by Andrew Balmford.

Figure 15. Dusti Becker and children from El Suspiro catch up on their field notes during the annual Christmas Bird Count. Photo by Andrew Balmford.

Figure 16. Bambil: what happens when all the fog forest upstream is gone. Photo by Andrew Balmford.

Figure 17. Burning early to burn less: early-dry season fires on the western Arnhem Land Plateau. These usually go out overnight as dew settles, leaving firebreaks that stop the advance of late-dry-season wildfires. Photo by Peter Cooke, Warddeken Land Management.

Figure 18. Restoring a mine pit after bauxite extraction. In the distance, intact jarrah forest grows up to the edge of the pit wall. Photo by Andrew Balmford.

Figure 19. *Macrozamia* seedlings being hand reared at Alcoa nursery. Since this photo was taken, Alcoa scientists have discovered they can get these cycads to germinate directly at the mine site, provided the seeds are carefully prepared and then buried in a process that mimics their natural passage through the gut of an emu. Photo by Andrew Balmford.

Figure 20. Western pygmy possum, with zoologist Mike Craig's finger for scale. Photo by Andrew Balmford.

Figure 21. Pole-and-line fishing albacore from the rack of an AAFA vessel. Photo by the American Albacore Fisheries Association.

Figure 22. Sustainability in a jar: AAFA albacore, certified sustainable by the Marine Stewardship Council. Photo by Andrew Balmford.

Figure 23. Part of the Ely Wildspace, with the city's Norman cathedral in the background. Photo by Kevin Smith, Ely Wildspace.

Figure 24. Some took the campaign to stop a marina being built in the Ely Wildspace to an extraordinary degree. The campaign's bittern image by Michael Edwards; tattoo by Fabio Giovannoni; arm by Russ Malster. Photo by Andrew Balmford.

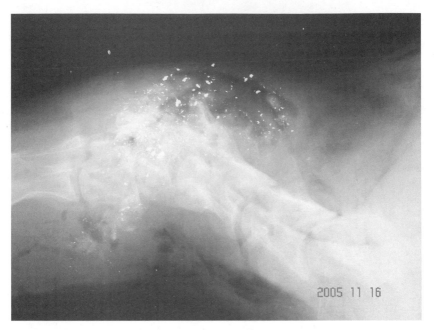

Figure 25. X-ray image of lead fragments scattered in the neck muscle of a mule deer. Showing hunters pictures like this has persuaded many that switching to copper bullets is better for their own health as well as that of endangered condors. Photo by Chris Parish, Peregrine Fund.

THE GREENING OF A GIANT

So how far can efforts to link conservation to mainstream economic activity go? Is big business getting involved, and if so, why? In the unique jarrah forest of southwestern Australia I find out how and why the largest aluminum mine in the world has also become the greenest, carrying out cutting-edge research in rare-plant propagation and partnering government in a statewide fox and cat eradication program involving a locally derived poison, an Italian sausage machine, and precision bombing with chipolatas. Why all this effort, way in excess of legal obligations? Because the company believes that, regardless of formal contracts, their real license to operate a highly profitable concession into the future depends on long-term public approval.

For a moment it's as though I'm living out a nightmare. Barren red earth stretches hundreds of meters in every direction. The engine roar from the biggest vehicle I've ever seen competes with red dust to fill the hot, dry summer air. In the far distance I can make out the abrupt wall of the mine pit, as tall as a house, and just above it, the fragile, olive-green edge of a unique forest, called jarrah, that only a few months ago reached all the way to where I'm standing. It's an ecological Armageddon. And now a wiry bloke with a biker's handlebar mustache and the body language of an extra from a Hollywood bar brawl is clambering down from his 110-tonne bull-dozer to have a word with me...

But here in the largest bauxite mine on the planet—Alcoa's operation in Huntly, Western Australia—things are not quite what they seem, and Bradley "Buzzard" Ambrosius is not exactly your ordinary miner.

Shock and Ore

Geology dictates that there is no way of extracting bauxite—the raw material for making aluminum—without digging a very large hole in the ground. In the low-lying Darling Range an hour's drive southeast of Perth, the bauxite lies in a shallow layer 50 centimeters to 5 meters below ground, and open-cast mining starts with pilot drilling to find suitable ore bodies where the bauxite is concentrated. Miners then clear away the jarrah forest and the soil from an area up to half a kilometer across, blast through a concrete-like layer called caprock, and finally use excavators and trucks bigger than houses to dig up the bauxite and haul it off to a crushing plant, from where a conveyor belt transports it to a refinery 25 kilometers away. That's when what most of us think of as mining is over. But it's when Buzzard's work begins.

That's because he's one of more than 80 people at the Huntly mine charged with trying to put things back how they were, on a massive scale: not just the soil, but the trees, the myriad other creatures of the jarrah system, and all the ecological threads that link them together. Buzzard and the other bulldozer operators begin the rehabilitation process. "The mining side of it is straight in, straight out, but the rehab side of it is phenomenal. We start with smoothing the pit edges, knocking all the high parts down with the dozers, and burying the big rocks," he explains. Once they've finished converting the steep pit walls into gently sloping banks, the dozers work as plows, dragging vast, anchor-shaped tines beneath the soil to break up the compacted pit floor so that water can drain and tree roots can penetrate.

Next, flat-bellied vehicles larger than London buses trundle over from newly cleared pits nearby to deposit two layers of soil, with unlikely precision: first, around 50 centimeters of the deeper so-called overburden, and then between 10 and 15 centimeters of fresh, seed-rich topsoil. Large logs and rocks are put back to provide starter homes for recolonizing animals. And then the bulldozers return to plow along contour lines while sowing many thousands of carefully selected seeds using computer-driven seeding machines. Hundreds more plants are put in by hand, fertilizer is dropped on the site from a helicopter, and finally, if all the procedures have been followed correctly and the winter rains arrive on time, a new forest, with a ground level 4 or 5 meters lower than its predecessor but with a remarkably similar set of species, begins to regrow.

As he tells me all this, it's obvious Buzzard gets enormous satisfaction from his work. "When you see a nice finished pit, it's good when it's all seeded, and it all takes off." Royce Edwards, a jovial and straight-talking ex-policeman who only intended to work for Alcoa for 2 years but has stayed there for 26, smiles as he expresses the same sentiments: "I pass places on my way to work which I mined—but now I can't tell I've been there. That's the bit I like." It's clear this pride in trying to do the right thing cuts right across the workforce. Glen Ainsworth, the manager overseeing the rehabilitation work at Huntly, adds, "There was a time when a dozer operator used to be just a dozer operator—paid to come to work and not to think. But now they're responsible for putting things back the way they were before. And a lot of the guys are bloody proud of what they do."

Nature Up a Gum Tree

Many of us worry about the environmental impacts of mining. Poor practices have caused widespread damage to valuable habitats, contaminated water supplies, made healthy people sick, and left a lasting legacy of mistrust. Much harm continues to be done. But unless society as a whole performs a collective U-turn and decides it can live without its products—from cars to drink cans, from fridges to mobile phones—mineral extraction will continue; indeed, with rapidly escalating demand for raw materials from countries like China and India, mining is growing fast.

Mining is especially significant for conservationists, because a great deal of it happens in or near to places of high value for conservation. By comparing global maps of mine concessions and priority areas for conservation, the Washington-based World Resources Institute worked out that more than a quarter of all mine sites overlap with or are within just 10 kilometers of a "strictly protected area."[1] Even more starkly, almost three-quarters are in areas identified as being globally important for conservation. To some degree this is simply because nature reserves and other conservation priorities cover quite a high proportion of the world's land surface, but there's more to this pattern than chance alone. Mining activities are often concentrated in areas with unusual geology, where uncommon processes of rock formation or extremely long exposure to weathering have made scarce minerals accessible. But the effects of these same phenomena on the makeup of soils mean that these places also support distinctive and often species-rich communities of plants and animals, many of which are found nowhere else.

The ancient landscapes of Western Australia are a good example. Thou-

1. Strictly protected areas are those reserves falling in categories I–IV in IUCN's classification system. These include most of the highest priority reserves, and they receive strong protection on paper (although not necessarily on the ground). Most national parks, for example (like Yosemite or Serengeti), are category II protected areas—although those in the UK, which receive much less protection and comprise mainly farmland, are in category V.

sands of millions of years of erosion of the oldest rocks on the earth's oldest continent have left the region exceptionally rich in workable mineral deposits. As a consequence, most of the state's wealth now comes from mining—in particular for iron, gold, nickel, and aluminum; Rio Tinto, one of the world's largest mining multinationals, earns 40 percent of its total global revenue just from its operations in Western Australia. The state is also extremely rich botanically. The official tally stands at well over 11,000 species, but very many have still to be described.

Roughly half of this plant diversity is concentrated in the southwest corner, where kwongan, the antipodean equivalent of the South African fynbos, is to be found. As in other parts of the world that experience what geographers term a mediterranean climate (with hot, dry summers and warm, wet winters), the vegetation here consists mostly of heaths and scrublands—tough-leaved, shrubby plants adapted to seasonal drought, fires, and infertile soils, yet paradoxically highly diverse. But in the middle of these heathlands there's a forest. The jarrah eucalypt forest: the only forest anywhere in the world able to withstand the harshness of a mediterranean climate.

Jarrah trees are impressive beasts. They can reach as high as a blue whale is long, grow trunks wider than a man is tall, and live for several hundred years. Yet they do all this in a part of the world where there may be virtually no rain for 6 months and where summertime temperatures frequently top 30°C. The answer to this puzzle lies underground—in the water-absorbing clay soils that jarrah trees penetrate with 50-meter-deep root systems, enabling them to continue harvesting winter rainfall long into the summer drought. Add the species' capacity to survive fires and resist termite attacks, and jarrah can out-compete just about all other trees in the Darling Range.

There are a few other common trees and shrubs in the jarrah system— the marri (another large eucalypt), the flamboyantly flowered bull banksia (a relative of South Africa's proteas), the snottygobble (elegant and somewhat delicate despite its prosaic name), and the charismatic grass tree (looking like an oversize, bright green version of a 1970s fiber-optic lamp). But the real diversity of the flora is at ground level, where some 800 species—of banksias and peas, sedges and orchids, myrtles and bottlebrushes—make the jarrah forest one of the richest anywhere outside the tropics.

The low fertility of the ancient soils of the Darling Range means that unlike the woodlands to the east and the giant forests of karri trees in wetter hills to the south, most of the jarrah forest has been spared from clearance for farming. Instead, early European settlers used the forest for logging. Jarrah timber is finely grained, deep red in color, and virtually impregnable to termites and rot. As a result, it has been in demand for almost two centuries for cabinet making, flooring, and house construction, as well as for outdoor uses like fenceposts and telegraph poles. Western Australia's first recorded export, in 1836, was of 5 tonnes of jarrah railway sleepers bound for England; some London streets to this day are paved with jarrah cobbles.

As markets grew, logging rates increased. The jarrah forest did regenerate, but with the big trees gone, its appearance changed, with denser stands of short, young jarrahs, and a thicker layer of banksias and mop-haired grass trees underneath. By the 1970s, less than one-twentieth of the old-growth forest in the main northern block remained—but by then, public attention was starting to focus on two other threats to the system.

One was the expansion of mining in the Darling hills, and the clear-felling of the forest that went with it. Most of the deep clay soils in which jarrah trees thrive are covered with a layer of bauxite, and the combination of low levels of contaminants together with relatively easy access meant mining it to make aluminum was highly profitable. Bauxite mining leases were granted over almost all of the northern portion of the jarrah forest, an area roughly 50 kilometers wide and 200 kilometers long that is also the principal catchment for most of Perth's drinking water. After a series of feasibility studies in the 1960s, large-scale mining operations began in earnest.

The second major worry was the emergence of a devastating disease. *Phytophthora cinnamomi* is a microscopic, fungus-like organism that infects the roots of plants, disrupting their ability to transport water and nutrients and eventually killing them. No one yet knows how this alien invader got to Australia, but the most likely suggestion is it arrived in the roots of an infected plant brought in by nineteenth century European settlers. Although some jarrah trees started dying in the 1920s, the link to *Phytophthora* wasn't established for another 40 years—by which time road-builders, ironically trying to avoid damage to healthy forests by only excavating gravel from areas where jarrah had died, had inadvertently spread what became known as jarrah dieback disease far and wide. As the rot deprived them of water, infected trees shed their leaves, exposing eerily bare crowns and gaunt side-branches in an otherwise evergreen forest. Worse still, *Phytophthora* (which means "plant-destroyer" in Greek) turned out to have broad tastes and was capable of destroying not just the jarrah but around 40 percent of all the plant species in the system, as well. Dieback was emerging as a threat to the entire forest, and because *Phytophthora* spores are transported in the soil, many believed that open-cast bauxite mining could only make a bad situation worse.

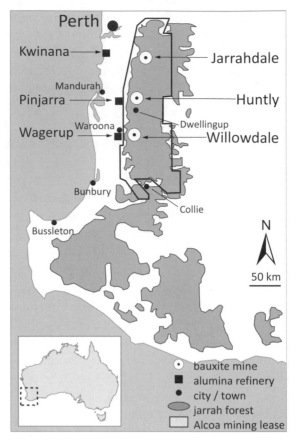

Map 7. Alcoa's bauxite lease in Western Australia covers most of the northern portion of jarrah forest. Map based on information in Koch 2006. Cartography by Ruth Swetnam.

With hindsight, however, the jarrah forest had one stroke of good luck in all this—that the company in charge of excavating much of the bauxite reserves on which it grew was Alcoa.

Taking the Long View

In the early days, Alcoa did exactly what the law required in terms of restoring the forest it cleared—which was very little indeed. Under its 1961 agreement with the state of Western Australia—which granted it an exclusive, 84-year lease to mine ore bodies across an area the size of Connecticut—Alcoa was simply asked to return the topsoil to the compacted floor of each mine pit and leave the rehabilitation work to the Forests Department. With limited experience of growing forests from scratch rather than cutting them down, and being concerned about the susceptibility of jarrah to dieback disease, the foresters instead planted up the mined areas with rows of familiar but nonnative pines. No fertilizers were used to speed up establishment, and with their roots unable to push through the compressed ground left after mining, the young trees reached only a couple of meters in height before toppling over in the first strong wind, like so many matchsticks. Rehabilitation wasn't working.

Worried about the impact that the sight of unhealed mining scars and fears of contaminated water supplies might have on public opinion, Alcoa decided to act—to do more than the law required. In 1971, it took on its first environmental officer, an agricultural science graduate from London named George White. Catching up with George, twinkle eyed and still with a cockney edge to his voice nearly 40 years later, he tells me that at first the company had no idea what it was letting itself in for. "When the manager of mines interviewed me he said, 'We're not sure it's going to be a full-time job, but in your spare time you can always help out in the refineries.'" But George had bigger ambitions. "I took the approach that there was already a groundswell of public criticism. I started out life wanting to be a botanist, and so when I looked at what was happening to the jarrah forest, I thought 'This isn't right. We have to do something about this.' I put it to Alcoa that what we really needed to do, if we were going to be expanding for 3 or 4 decades, was to think seriously not just about meeting current expectations, but about anticipating the likely needs and aspirations of the community in 20 years' time. We had to start taking a long view of this."

Alcoa listened George's argument about anticipating public opinion, and invested in his vision. When he explained to his managing director that the problem with trees falling over was compacted soil, he was told, "Go to Perth and get the biggest bulldozer in the shop." More resources followed. Within 5 years, George had around 30 people working with him, in nearby universities as well as at Alcoa. "We hired hydrologists so we could understand the effects of mining on groundwater. We set up a rehab group responsible for getting the vegetation right and an animal man to look into the animals. I took on a microbiologist to work on dieback disease. And we hired a botanist whose job it was to understand all of the flora in the Darling Range, so we could put more of the native species back."

Research led to a steady stream of innovations. Pit walls were reshaped into shallow slopes to fit in with the surrounding land. Alien pines were replaced with eucalypts (albeit from eastern Australia) and even some experimental jarrah plantings. An Alcoa researcher invented an anchor-shaped plow blade that enabled bulldozers to break up the compacted mine floor and so facilitate decent root penetration. And the new trees were given a nutrient boost with fertilizers and the addition to the seed mix of native species from the forest understory, many of which soak up nitrogen from the atmosphere. As George explains, "We looked at just about every option for doing it better. There was never any real question of whether we could have the money—the company were totally and absolutely committed to doing the right thing. It just took us a while to work out what the right thing actually was."

But despite the improvements on the ground, public perceptions of Alcoa's operations were slow to shift, and by the late 1970s the company was facing a wave of protests against its application to expand capacity by building a new refinery. Environmental activists chained themselves to bulldozers, green-minded politicians railed against Alcoa in TV interviews, and the parent company faced a class-action lawsuit in the United States. In due course, George White's gamble of trying to anticipate years ahead of time the questions society would ask of it paid off, and the answers his group provided helped Alcoa to win the court case and the enquiry over the refinery. But the company still had work to do winning the hearts and minds of its public.

During the 1980s George White's successor, Barry Carbon, brought market researchers and experimental psychologists into the environmental team and embarked on a serious PR campaign. Extending work George had begun, tens of thousands of visitors were given tours to see the mining and rehab work firsthand. Most left surprised at how little of the overall forest was being cleared[2] and impressed by the company's restoration efforts. The company also applied its hydrological and horticultural know-how to help wheat-belt farmers hundreds of kilometers to the east tackle growing problems with saline soils. And Alcoa's nursery—already busy producing plants for restoration—gave away more than 2 million trees in an effort to encourage local people to grow native species in their gardens. Progress continued on site too. Alcoa's scientists and collaborators began to unravel the complex biology of dieback disease and slow down its spread. There were further improvements to restoration techniques. And the company started a process of identifying the most valuable areas for conservation within its lease, and excluding them from mining even when they contained viable bauxite deposits.

---

2. Because the mining is focused on ore bodies where the bauxite is concentrated, over its entire lifetime Alcoa's Huntly operation will only clear 4 percent of the forest.

Fast-forward another 20 years, and it's clear that these twin investments in cutting-edge restoration and in communication have paid off. As public opposition waned, Alcoa's production rose; by the turn of this century its operations in the Darling Range were supplying one-seventh of the world's alumina,[3] employing over 4,000 workers, and contributing more than AU$1 billion each year to the Western Australian economy. But the company was also winning environmental plaudits from around the world—a Golden Gecko for environmental excellence from the state government, a model project award from the U.S.-based Society for Ecological Restoration, and most prestigious of all, the 1990 inclusion (to match the company's Fortune 500 status) on the United Nations Environment Programme's Global 500 Roll of Honour—still the only mining company to receive the award.

Seeds of Recovery

Looking around a 15-year-old plot of rehabilitated jarrah, it's not hard to see why Alcoa's restoration work is rated as the very best in the world. The company has confined its mining to parts of the forest that have already had their bigger trees removed at least once. When comparing the rehabilitated area to the unmined but logged forest next door, it's difficult to spot the difference. John Gardner, urbane and energetic head of Alcoa's current environmental operation, is on hand to help me out. Both pieces of forest are dominated by the dense olive-green canopy and stringy gray barks of tall jarrah trees, interspersed with a few marris. In each, the understory is rich in unfamiliar bushes and shrubs, and the floor carpeted with leaf litter, logs, and fallen branches. But with John's practiced eye, I can begin to make out slight differences. The jarrah on the rehab site is slightly shorter, more evenly sized, and a bit more densely packed. The slow-growing grass trees are still tiny, and the dieback-sensitive banksias near-absent from what, only 180 months ago, was an open-cast mine. The restored forest is not perfect. But it is extremely impressive. And it's taken a great deal of painstaking trial and error.

More than 100 honors, master's, and Ph.D. students have carried out research projects aimed at improving Alcoa's restoration efforts. One of the main thrusts—spearheaded by John Gardner—has been fighting dieback disease. Back in his office, he explains the importance of this work: "When I

3. Alumina—or aluminum oxide—is the compound derived from bauxite from which aluminum itself is then extracted.

started in 1978 the prediction was that for every hectare of forest we cleared for mining, we would infest between 1 and 4 more with the disease. Dieback was a real barrier to our aim of putting back the jarrah forest." To shed light on how the disease worked, John and his team worked with other scientists to look at where exactly the pathogens set up home inside their hosts. To better understand how the disease spread, they took thousands of samples from the forest, the mines and the mine roads, and screened them for traces of spores. And to see if they could hold the disease in check, they conducted field trials and tested whether they could keep newly planted trees free from infection.

Little by little they pieced together much of what is now known about dieback and how to control it. They worked out protocols for mapping infested areas (not too difficult—just look for the dead plants), and then devised stringent quarantine procedures. Nowadays all soil from infested areas is kept strictly separate from uninfested soil, all mine vehicles leaving infested areas (from pick-ups to 200-tonne trucks) have to be hosed down in a gigantic chlorinated car wash, and road edges are specially landscaped to stop potentially contaminated water running off into the forest. The hygiene measures work. Exhaustive monitoring along almost 170 kilometers of the edges of mined areas found the infection was spreading along a combined front totaling just 80 meters; each hectare of mining was on average contaminating less than one-thousandth of a hectare more.

John's team also found out that the mining process itself was inadvertently making life difficult for *Phytophthora*. The disease does best in soils that are warm and wet—precisely the conditions created by the relatively impervious layer of caprock lying just under the jarrah forest. But mining smashes up the caprock, takes it away with the rest of the bauxite, and turns it into baking foil. This discovery solved a nagging mystery—why jarrah (which was so susceptible to dieback that foresters wouldn't plant it at all) had survived for many years when planted experimentally in infested areas. The answer: freed from the dampness previously sealed in by the caprock, jarrah wasn't actually all that susceptible. So even in infested areas, mining and subsequent rehabilitation could help stop the disease from flourishing. John says these findings energized Alcoa's wider restoration ambitions: "Once we knew we could put jarrah back, it was worth taking on the whole of the rest of the forest."

This was no mean undertaking—just in the parts of the forest where Alcoa is mining there are around 300 plant species—but research again helped. Germination from returned soil was far and away the most important source of new plants, but when investigators looked more closely at the bank of seeds in the soil they found they were concentrated in just the top 5 or 10 centimeters, and died off quickly when soil was stockpiled; they also found seed densities were twice as high in summer as in winter. So the rehab teams started separating the topmost soil from the rest of the over-burden, cutting out stockpiling wherever possible by immediately putting topsoil removed from one pit onto a freshly rehabilitated pit nearby, and carrying out as much of this work as they could over the summer.

Despite these advances, many species were not recolonizing from the soil seedbank, so Alcoa worked at improving its direct seeding. Nowadays the company pays commercial collectors to gather over 4 tonnes of seed from around 100 target species each year. To limit the depletion of wild popula-tions, efforts are focused on areas that are about to be cleared for mining; to avoid losing the plants' genetic adaptations to local conditions, all seeds are sourced from within 20 kilometers of where they will be sown.

At the company's nursery, John Koch, a tall and softly spoken bota-nist, shows me around the low temperature seed store. We first negotiate chunky padlocks—some seeds are so hard to collect they're literally worth more than their weight in gold. Once inside, sacks and bins and canisters overflow with seeds of every size and shape imaginable—some winged, or keeled, others encased in large nuts, still others tinier than grains of sand. He tells me the Latin binomial of each species as if he were introducing his extended family. In a sorting room nearby, technicians armed with weigh-ing balances make up seed mixtures to precise recipes, for broadcasting onto fresh topsoil once the summer arrives.

Getting smart with soil and seeds helped Alcoa reach an important, self-imposed target: raising the number of plant species in 15 month-old rehab plots from 65 percent to 80 percent of the number in unmined for-est plots over the course of just 5 years. Still not satisfied, they voluntarily raised the bar again, and in 1995 gave themselves just 5 years more to reach 100 percent—to make their restored forests on average every bit as rich as the forests that preceded them. Getting there would require more smart thinking, and further partnerships with the outside world.

Noah's Park

I'm standing in a state-of-the-art tissue-culture lab surrounded by thou-sands of glass vials arrayed under bright fluorescent lighting, and an infec-tiously enthusiastic scientist is passing me a jar whose inhabitant should by rights be extinct. We're in the new research facilities of Kings Park and

Botanic Garden in Perth, and the plant in question is the scarlet snakebush, a Western Australian native now extinct in the wild but, thanks to high-tech cloning, still clinging on as a captive population (albeit one whose individuals are all genetically identical). My engaging host is Kingsley Dixon, director of the lab, and plant propagation entrepreneur *par excellence*. Combining a fierce-eyed passion for all things botanical with the persuasive patter of a used-car salesman, Kingsley is just the person I'd want fighting my corner if I was a plant teetering on the brink of extinction. A twenty-first-century Noah with a lab bench for an ark.

One staggeringly rare species that's safer thanks to the Kings Park team is the Dwellingup mallee, a eucalypt known from just a single wild plant growing on an ancient granite outcrop on the Alcoa concession. "Being the good corporate citizens that they are, Alcoa were committed to ensuring they would have no negative impact on this only surviving individual," Kingsley explains. So with their support, Kingsley and his colleagues first worked out how to clone the plant—a task that proved far from straightforward but eventually allowed them to produce a priceless back-up population of a few hundred plants. Next, they conducted genetic fingerprinting to discover more about the species' origins and found out it was in fact a natural hybrid, produced thousands of years ago when unusual climatic conditions allowed its two parent species—one arid-adapted and one riverine—to grow in close proximity. "We now use the plant as an icon for our outreach work in schools, to teach children a bit about conservation, genetics, the mining industry, ancient climates, and about how the species in our landscapes are precious because they may have been around as individuals for extraordinarily long periods of time."

Kingsley has also played a pivotal role in helping Alcoa edge toward their 100 percent target by getting many more common but stubborn jarrah forest species back onto its rehabilitation sites. While on sabbatical in South Africa in the early 1990s, he met up with fellow botanists Neville Brown and Johannes de Lange, who a few years previously had made a major breakthrough in getting rare Cape plants to germinate. They had known that the seeds of many fynbos species only broke dormancy after fire (which makes good evolutionary sense, as fires provide gaps for new plants to exploit), but despite treating these seeds with heat and ash, they were unable to persuade them to germinate in the lab. Then one day, during an experimental burn in the field, the wind changed direction, and blew smoke straight in Johannes's face. His eyes started watering, and suddenly the penny dropped: if smoke could cause a physiological reaction in him, perhaps it was capable of triggering a physiological response in a seed.

Kingsley takes up the story: "We'd all looked at heat and ash but we had overlooked the third and most obvious component: the stuff that gets in your eyes. So next they went to a site where they were trying to save a critically endangered species with a global distribution the size of this table, set up a bellows and a small fire under a tarpaulin, and the magic began."

Neville and Johannes quickly discovered that exposure to smoke—or even to water through which smoke had been bubbled—provided the vital cue needed to coax dozens of reluctant fynbos seeds into life. Kingsley describes coming face-to-face a little while later with thousands of smoke-triggered seedlings as "one of those eureka moments. We'd been working for years trying to crack deep dormancy in a whole range of Australian species, and here was the answer. I was so excited I stayed awake on the plane all the way back from South Africa. And then after a year or two of trying to get it to work on Australian species, we started to see this eruption of germination. Glorious wildflowers, charismatic species that none of us could ever grow, were coming up like cress!"

Encouraged by Alcoa, the Kings Park team soon began smoke experiments on seeds from the jarrah forest. The results were striking. In some trials, the number of species germinating increased by 30 percent, and the number of seedlings rose threefold. Alcoa was swift to apply the findings, smoke treating one-third of all the species in its seed mixes before sowing them on the mine pits. A few other fire-adapted seeds—especially some of the acacias—responded to heat instead, and so nowadays are parboiled before sowing. For a handful of other species—snottygobble included—breaking seed dormancy is still proving elusive. But for the great majority, provided they can get the seeds, Alcoa can now persuade them to grow.

The last big hurdle to closing in on Alcoa's 100 percent target has turned out to be a few dozen "recalcitrant"[4] species that barely produce seed at all. Once they are established, these plants—mostly grasslike species such as sedges, rushes, and grass trees—tend to spread vegetatively (through underground tubers, bulbs, and the like) rather than sexually (by flowering). This makes them excellent at recovering quickly from fires and resprouting after kangaroo grazing, which in turn means they are ecologically very important: they confer resilience on the system. But it also means that if the plants are removed completely (as they are at the start of mining), it's very hard to reestablish them.

---

4. Alcoa defines "recalcitrant" differently than most botanists, who use it to describe seeds that don't survive desiccation or freezing and so are difficult to store for long periods in artificial seed banks.

Alcoa's solution has been multipronged. By pampering them with the horticultural equivalent of soft lighting and mood music, it manages to entice a few recalcitrant species to flower and set seed in the nursery. Others are the focus of an industrial-scale cuttings operation: tens of thousands of shoots are removed from wild plants and brought back to the nursery for trimming, dipping in fungicide and growth hormones, and potting up. And the most difficult species of all are propagated in the company's own AU$3 million tissue-culture lab, where a repeated cycle of taking several cuttings from a single plant, growing these up on sterile agar jelly, and then subdividing them into more cuttings, generates over 100,000 plantlets each year. In a typical season up to a quarter of a million recalcitrant plants are produced altogether (at around AU$5 each), and then hand-planted in the rehab sites.

Like Buzzard, the nursery workers take well-earned pride in what they're doing. I catch up with Maureen Howard, who's lived in the area her entire life, as she carefully rinses and trims a large pile of cuttings: "I love the forest, and now it's coming back. That's a good feeling."

The fruits of Maureen's hard work are on show all around me when I go out a few hours later to a 2-year-old restoration site. Vigorous sedges and rushes are pushing tender shoots through the biodegradable mesh bags that have protected them thus far from becoming kangaroo fodder. Knee-high jarrahs and marris are starting to accelerate upward. Log piles have been colonized by spiders and ants. The great scar of the mine pit is beginning to heal over.

Nowadays Alcoa's rehabilitation work is so efficient that the company restores ex-pits as fast as it creates new ones—at the rate of about 550 hectares (or around 700 football pitches) a year. In line with its vision of mining being only a transient land use, Alcoa has completely closed one mine area and started handing the rehabilitated forest back to the state authorities. Even more importantly, as of 2001 the company reached its new target and recorded as many (in fact slightly more) plant species in its new rehab areas as in the average patch of unmined forest.

But Alcoa's concerns aren't limited just to plants, and alongside its botanical monitoring, it has students and staff looking at whether restoration works for a host of other organisms, from termites, bugs and springtails, to fungi, mites, and even soil microbes. These kinds of creatures are in some ways less amenable to manipulation than plants, but they play vital roles in ecosystem functioning—from recycling nutrients to turning over the soil—so getting them back is crucial to reestablishing healthy habitats.

An Aussie Bestiary

To learn more about the zoology of jarrah rehabilitation, I've come out on an unseasonably damp morning with Mike Craig from Murdoch University and his new Dutch research student, Maree Weerheim. Mike's running a project looking at how various post-restoration treatments of the forest—burning it or not, thinning out the jarrah to differing degrees—affect vertebrates like reptiles and marsupials. It's soon evident that he adores his job.

To find out much about small, shy vertebrates you have to trap them, and Mike's trapping regime is no-nonsense stuff. Each trapline has a 30-meter-long mesh drift-fence. Animals that encounter this scurry along it and then have a choice of where to spend the night—in 1 of 9 pitfall traps, 4 funnel traps, 4 aluminum traps baited with peanut butter and oats, or failing that, in a wire cage at the end. The next morning, Mike and Maree identify and measure the captives, before releasing them unharmed and moving to a trapline at another site.

Today, despite the rain, we've struck lucky: it's a bumper haul. A daylight-dazed, velvet-skinned gecko. A cat-sized bundle of grumpy gray fur that reveals itself to be a brushtail possum. And an impressively bulky lizard called a bobtail, whose head and tail both look like flattened pine cones and that lunges up open-mouthed to startle with its cobalt-blue tongue (like me, would-be predators are either baffled or alarmed—or both). Most of the catch consists of little skinks—lithe, striped lizards that almost swim through the leaf litter and come in confusing variety. Golden tails, red tails, tiny limbs, even absent limbs (all the better for swimming); 5 species in all. Explaining which is which and why, marveling at the details, Mike is in his element. But for me, the real jewels are two tiny, bead-eyed marsupials: a flower- and insect-eating western pygmy possum, which weighs less than its name in Scrabble tiles, and a ferocious shrew-like carnivore known as a mardo (or yellow-footed antechinus), a beast I've yearned to meet for almost 20 years, ever since learning about its extraordinary life history. At 11 months old, male mardos put so much effort into their one and only mating season that they die straight afterward. For the next 2 months the only males in the entire population are those inside their mothers (as embryos or pouched young). Coming face-to-face with my first mardo, I grin like a Cheshire cat.

Data already gathered shows that most vertebrate species in the jarrah forest do return to the rehab sites, although sometimes in smaller numbers than before. Mike's work will provide important additional insights into how Alcoa can boost populations, by fine-tuning habitat restoration so that the returnees can find enough food, and sufficient holes and crevices to shelter in. But it turns out that the biggest problem facing most of the marsupials—not just in mined areas but across the region—is nothing to do with habitat loss. It's a deadly duo of introduced predators: the European red fox and the domestic cat.

The Tale of the Flying Salamis

Cats arrived in Australia alongside European settlers almost 200 years ago; pining for the thrill of the chase, the British then introduced foxes some time in the 1860s. Between them these highly efficient predators have since caused continent-wide reductions of just about every native mammal larger than a mouse but smaller than a Jack Russell and that can't fly, burrow, or climb trees to escape. Eighteen of these medium-sized mammal species have already become extinct while dozens more have declined very substantially.

For a while, the southwestern forests seemed relatively immune—the cats stayed in the drier areas inland, and the impact of foxes was limited. But during the 1970s and '80s, populations of native mammals began plummeting even in their southwest strongholds. Circumstantial but convincing evidence pointed the finger at the fox. Fortunately, worried wildlife managers had a powerful weapon at their disposal, a potent poison formally called sodium monofluoroacetate but better known (from its catalog number) as 1080.

First discovered by Nazi chemists during World War II, 1080 has long been used for poisoning problem mammals, including introduced rabbits in Australia and pest possums in New Zealand. Its potential as a tool for conserving native Australian mammals came to light in the 1970s, when scientists worked out it was the same as a naturally occurring toxin found in over 30 species of plants called poison peas that come from the bottom left corner of the continent. Linked with this, they found that native southwestern mammals (long exposed to the poison in nature) are unusually resistant to 1080—pound for pound it takes 100 times as much to kill a mardo as a fox, and 1,000 times as much to kill a Western Australian brushtail possum.[5] This in turn meant that in the southwest at least it might be possible to put out 1080 in poison baits and kill nonnative predators without harm-

5. Interestingly, east Australian brushtails that have not evolved alongside poison peas are far more sensitive.

ing the locals—so that's exactly what the Department of Conservation and Land Management (CALM) started doing.

The results were impressive. Experimental trials conducted in the mid-1980s in isolated nature reserves out in the wheat belt led to marked recoveries in populations of two hard-hit marsupials: the dapper-looking black-footed rock wallaby, and a charming ochre-and-white-banded termite hunter called the numbat. Both were suspected victims of foxes (which had probably increased because a rise in myxomatosis meant farmers had stopped putting out 1080 baits to control rabbits). Hand-laying fox baits around the edges of the reserves confirmed suspicions: large numbers of foxes were killed and local populations of numbats and rock wallabies rose, even while other, nonbaited populations continued to freefall. The next question conservationists at CALM (and indeed Alcoa) wanted answers to was whether these isolated programs could be scaled up.

With financial backing from Alcoa, CALM embarked on Operation Foxglove, a large-scale trial to look at the feasibility and ideal frequency of fox-baiting across a 5,500 square kilometer swath of jarrah forest. The area was clearly too big for laying baits by hand, so instead an aircraft fitted with a dedicated on-board computer and global positioning system was commissioned to fly precise transect lines, 1 kilometer apart, dropping one poisoned chunk of dried kangaroo meat every 200 meters. To avoid killing household pets, the GPS and computer were programmed to tell the baiting machine to stop the drops whenever the plane was within half a kilometer of a private property.

Areas were baited 2, 4, or 6 times a year, or left as unbaited controls, and the trials produced encouraging results. Compared with untreated control areas, baiting led to falling fox numbers and increasing populations of brushtails, as well as quendas, woylies, and chuditch (a.k.a. bandicoots, small wallabies, and spotty marsupial cats, respectively). More frequent baiting produced stronger responses.

With large-scale but selective poisoning now a practical possibility, in 1996 CALM expanded Operation Foxglove into an even more ambitious program called Western Shield. Billed as the biggest wildlife conservation intervention ever undertaken in Australia and again partly funded by Alcoa, it extended predator control across 32,000 square kilometers of nature reserves and other prime habitat in Western Australia (including the jarrah forest). The scale of the operation is vast: one complete treatment requires 150,000 baits and 55,000 kilometers of flying. With rebaiting conducted every 3 months, the program's Beechcraft Baron plane is in the air for 8 months a year, traveling the equivalent of 5 times round the world.

Western Shield also restocks baited areas with small populations of some of the native animals they've lost. By 2003, the program had carried out more than 80 relocations involving 23 threatened species. Two-thirds of those running long enough to be assessed were judged successful. More significantly, two threatened species (the quenda and a wallaby called the tammar) benefited sufficiently to be taken off the threatened species list, while two others (the chuditch and the numbat) had their threat status downgraded from Endangered to Vulnerable.

However, not all native species have fared so well under Western Shield. Some (like the woylie) have increased but then declined once more. Others have not responded at all. It seems they're threatened by problems other than foxes: disease is one possibility, another is predation by feral cats. Cats are much pickier than foxes about what they eat and won't take dried bait. Not only that, but somewhat ironically, their southwestern population appears to have grown to fill the predator vacuum created by fox eradication. The answer CALM came up with? A AU$60,000 sausage-making machine.

Armed with their new piece of kit, CALM's researchers began developing designer sausages to poison even the most discriminating pests. A new 85-gram dried kangaroo salami called Probait, aimed at foxes. A bigger version for feral dogs. And a 25-gram soft kangaroo chipolata called Eradicat (available commercially as Puss-Off) that, thanks to help from a commercial pet food manufacturer, has just the right blend of choice herbs and spices to make it irresistible to feral cats. Altogether CALM now makes over 2 million sausages a year, all laced with 1080 and air-dropped with computer-guided precision across an area the size of Wales and Northern Ireland combined.

Digging Deep

It's too early to know yet whether aerial bombardment with poisonous chipolatas will make a real dent in the numbers of feral felines, but the verdict on most of the other restoration work Alcoa is involved with seems fairly clear. When I came out to Australia to see the Huntly operation firsthand I expected to encounter at least a few serious criticisms of Alcoa's rehabilitation work there—to meet one or two well-informed objectors who question the reality behind the corporate brochures. But even hardened conservation campaigners who've fought to close down mining operations elsewhere have real respect for what Alcoa is doing in the Darling Range.

Steve Hopper, former director of Kings Park and now in charge of the

Royal Botanic Gardens at Kew, tells me, "They've done some amazing things for a mining company. Compared with their colleagues, Alcoa has made a genuine attempt to invest in the science, trial possible solutions, and then scale up." Kingsley Dixon, who campaigned with Steve and others to prevent open-cast coal mining in what became Mount Lesueur National Park to the north of Perth, says, "As regrettable as losing 550 hectares of prime hardwood forest each year is, we got lucky. We've got the best company in the world, in terms of their ethic in restoration and management."

Even now, there are plenty of ways Alcoa wants to improve its jarrah rehabilitation. Saplings are still packed a little too close together. Coming up with the right thinning and burning prescriptions would also help return structural complexity to the forest. Current cloning techniques are too expensive to put back some of the previously abundant recalcitrant species—like the sedges and rushes—in sufficient numbers, so Kingsley and his team are busy piloting new, cheaper tissue culture methods. He's also keen to make the use of wild-collected seed more efficient, so that seed harvesting is less of a drain on wild populations. And of course there's always the cat problem.

But unless there's a major change in the philosophy underlying the company's environmental work—striving to do the right thing and raising standards ahead of public and legal expectations—you can bet on Alcoa meeting these challenges. As George White reflects, "They haven't always got it right, but they've always wanted to get it right, and that's the important thing. If you want get things right enough, you'll do it eventually."

A greater challenge facing Alcoa is how to spread the experience and attitudes built up in Western Australia right across the company's global operations. Alcoa works in over 40 countries and mines bauxite in five. In most places, national environmental restrictions are far less stringent than in Australia; in some countries there are barely any regulations at all, and governments are not necessarily keen for companies to forego taxable profits by doing more than the law requires. Yet for John Gardner, now charged with raising environmental performance company-wide, maintaining consistently high standards regardless of local expectations is very important; the company's antipodean achievements guarantee he has high-level support. "Getting our work in the jarrah forest put on UNEP's Global 500 Roll of Honour made everyone from here to New York puff their chest out. But they also realized they can't do that and at the same time know they're doing a crap job somewhere else. So that drives performance everywhere."

John and his group are now frequent visitors to Alcoa's bauxite mines in

Suriname and Guinea, advising on restoration techniques and helping to train up local rehab teams. Although it's slow work, for Huntly rehab manager Glen Ainsworth it's the most rewarding aspect of the job. "That was the big hitter for me—going to Guinea and helping people who wanted to do something but didn't know how. They used to be of a mindset that they were there to dig a hole, get paid, and get food—but not replace the environment they took away. But they see it now—so that's been a great thing for me. We're rehabilitating 250 hectares a year there now."

But progress in Guinea and Suriname notwithstanding, there's an even bigger challenge sitting in John's in-tray—one that extends beyond Alcoa altogether. It's the problem of how to bring the rest of the mining industry up toward Alcoa's standards. As co-chair of a biodiversity advisory group for the industry's International Council on Mining and Metals, it's an issue that takes up a lot of his time. There are a few examples of high-quality restoration being carried out by other companies: Kingsley tells me about a sand mining operation on the Perth coastal plain that is painstakingly putting back the banksia woodland it's removed to far higher standards than required by law.

Yet for Kingsley, while this and Alcoa's jarrah work are genuine bright lights, "they're but pinpricks in a dark tunnel. Nearly all the rest of the mining industry falls well short of doing a small fraction of what Alcoa is achieving." He talks of the destruction of extremely unusual vegetation at industry giant BHP Billiton's nickel mining site at Bandalup Hill—even before a botanical survey was completed, and even though the mine itself was uneconomic and closed soon afterward. Such behavior is the still the norm, he says. Operations like Alcoa's are the exception.

A lot of the problem lies in companies (and regulatory authorities) not taking restoration activities as seriously as they might, or doing so only at managerial level and not on the ground. But the other, more controversial side of the coin is the issue of whether some places are simply so important for conservation that they should be no-go areas for any mining whatsoever. The IUCN and others have called for mining to be stopped in strictly protected areas (though even this would not protect important areas that have not yet received legal recognition). Many mining companies, including some of the largest, disagree with this proposal. Some argue that even high-priority areas can be mined in principle, provided damage in one location is offset by help enhancing conservation efforts elsewhere. Yet who decides how much offset is enough? If we accept horse trading protected areas to make way for mining, why not for farming, or for golf courses? And what happens where species occur nowhere else? We would be horrified if a major multinational started mining beneath the Sistine Chapel. Surely, some natural treasures too should be untouchable.

Digging Deeper

The Alcoa story is a shining example of making conservation the business of business, and not just the narrow concern of a few conservation organizations. Many conservationists believe that such mainstreaming into the world of commerce (and government, and education . . .) is vital if we are to scale up from winning a few battles to not losing the war. Of course, how far other mining companies (or indeed any other members of the mainstream) end up following Alcoa on their remarkable journey comes down to motivation—so the last and most important question to ask about jarrah restoration is a simple one: Why?

Why is a world-class mining company smoking seeds, worrying about the germination of arcane plant species few of its shareholders have ever heard of, and pump priming statewide fox control? Why is it spending upwards of AU$35,000 rehabilitating every hectare it mines to a level far in excess of regulatory requirements? Why is it paying over AU$100,000 each year for SatNav salamis? And why is it invoking its competitors to be similarly conscientious? Why has the giant gone green?

For Alcoa, a large part of the reason is the belief that sound environmental management makes good business sense. As George White recognized over 30 years ago, the high profile of the company's bauxite mines—only an hour from the largest population center for almost 3,000 kilometers, and in the heart of that city's water catchment—makes its operations especially sensitive to public opinion. Although in a legal sense Alcoa has a license to mine bauxite for almost a century, in a practical sense its license to operate is contingent on the continuing goodwill of society at large. Should the company's performance fall seriously short of shifting public expectations, that right to mine would probably be withdrawn regardless of current legal agreements. Kingsley Dixon argues that a new mine that close to Perth would not now get the go-ahead to clear so much jarrah even on a one-off basis, let alone annually. Yet by continuously improving its rehabilitation work, Alcoa has kept up with society's increasing demands, and been allowed to expand a highly profitable mining operation.

There are other dimensions to the business case for conservation too. Some big customers actively prefer to buy aluminum from suppliers that look after the environment. Certain conservation practices, like returning topsoil directly rather than stockpiling it, simultaneously lower costs. And public support earned by doing a good job of restoration can give a PR boost right across the company—indeed, sound practice can help the reputation of mining as a whole.

So while top-class rehab isn't cheap, it can give a good return on investment. In fact, in an operation where a replacement tire for a giant truck costs AU$25,000, restoration work adds just 50¢ to the AU$7 cost of extracting a tonne of bauxite. But all that said, my impression from my time with Alcoa is that their main motivation for decent environmental stewardship comes from something more profound than just concern for the bottom line: it stems from an ethical desire to do the right thing that cuts through the organization, from top to bottom.

On the one hand, passionate and prescient champions of the environment—people like George White, Barry Carbon, and John Gardner—have been given strong support from above. Barry Carbon believes that the key at the highest level is not the business case—it simply boils down to whether the people that run a company care about whether they're liked by society. Most do, he argues, so the trick lies in ensuring environmental decisions are taken up at board level, by people who have the freedom to look beyond the bottom line. With their support, environmental managers can then set about enabling a caring ethos across the workforce.

In the case of Alcoa, the breadth of that employee commitment is impressive, and it's also understandable. One manager explains that for many of his workers the mine is their next-door neighbor, so what they do at work affects their lives very directly. If they make a mess, they have to live with it, and vice versa. Legacy is important too. Rehab group leader Royce Edwards explains, "We need to make this a good place for our children to grow up in. I want my kids to be able to come back here in 50 years and not be ashamed of what we've done. 99 percent of the people here want that."

For me, hearing these views from normal hardworking people is not only moving—it carries an important message. How far and how fast other companies follow Alcoa's lead will depend on many factors that determine the business case for conservation: the costs of looking after the environment, the regulatory framework, and the strength of pressure from campaigners and NGOs. But the opinions of employees and the value that a company places on retaining a caring workforce are also crucial.

So too—as I discover in my final journey, when I turn from the land to the sea—are the views of the consumers and companies that buy an organization's products. Conservation is primarily about ordinary people, and when they care enough, the extraordinary becomes possible.

FISHING FOR A FUTURE

The business case for conservation doesn't end with mining, and it can be shaped by all sorts of people—including readers of this book. Focusing on the unpromising example of the seas, where conservation problems seem at their most intractable, my last story shows how independent certification by the Marine Stewardship Council is enabling major retailers and conscientious consumers to encourage sustainable fishing practices in many different parts of the world.

For me, this final leg of my journey started a few years before I began work on this book, in the queue for the fresh fish counter at my local supermarket. I like shopping for fish. Not just because of the prospect of a good meal, but for the exotic spectacle of it—part biology lesson, part art exhibition. The myriad colors and shapes and life forms, the sheer strangeness of the creatures on display—animals from an unfamiliar world.

I scanned the bank of seafood, looking for what to buy for a dinner party. Thick red slabs of swordfish flesh laid out on white ice crystals; lurid yellow fillets of smoked haddock; sinister-looking monkfish with gaping, tooth-filled mouths; and regular ranks of tiger prawns, anatomically intricate, each one an exact replica of its neighbor. When I reached the front, I asked the assistant what he would recommend that came from a sustainably managed fishery. Why did I ask? Simple, really. Because the world's oceans are in a sorry state.

According to the Food and Agriculture Organization's annual assessment, widely perceived to be distinctly conservative in these matters, by 2008 31 percent of the world's fish stocks were "overexploited" or, worse still, "depleted," and 53 percent were "fully exploited . . . with no room for further expansion."[1] Just 15 percent were reckoned to be "moderately ex-

1. Food and Agriculture Organization of the United Nations, *The State of World Fisheries and Aquaculture* 2010, p. 8.

ploited" or "underexploited," and in only 1 percent of cases were fish populations "recovering from depletion." Put another way, four-fifths of all fisheries are being harvested right up to or beyond sustainable levels. In a world where people rely on fish for around one-sixth of their animal protein (with this figure rising to 50 percent in countries like Bangladesh and Indonesia) and where human populations are set to become half as large again in just a few decades, the current situation is a damning indictment of how we've managed our last great source of wild-caught food.

So, back to the supermarket queue and my quest for sustainably sourced supper. What could the assistant recommend? He asked me to repeat my query, then reassured me "Everything. All our fish is sustainably sourced, sir." No it's not, I replied, pointing to the line of orange roughy—deepwater fish that live cold and correspondingly slow lives half a kilometer or more beneath the waves. They can live for over a century, but because they don't breed until their twenties, and then do so in predictable spawning aggregations over sea-mounts, they've proved highly vulnerable to industrial overfishing, and are now considered by many to be threatened with extinction. "These fish aren't sustainably managed—but what have you got that is?" "No, sir, you're wrong," the salesman insisted wearily. "All our fish come from sustainable sources. So—what is it you want?" "Never mind," I said, and turned away. "Some people," the hapless assistant muttered to the next person in the queue (who just happened to be a close friend, and so passed on the compliment). Some people, indeed.

Plenty More Fish in the Sea?

Like charity, overfishing begins at home. In northern Europe, fishing was, until the eleventh century or so, a largely freshwater affair, centered on vast spawning migrations of fishes[2] like salmon, sturgeon, and shad, which spend most of their lives at sea but return to rivers to breed. But mediaeval overfishing, dam building, and land reclamation greatly reduced the annual influx of fishes, and attention switched to the coasts.

The rewards were rich. Callum Roberts's shocking account of the long story of overfishing, *The Unnatural History of the Sea*, describes ling, cod, and pollock remains found in Viking deposits from the north of Scotland

---

2. So what's the plural—fish or fishes? I'm reliably assured by fishy friends that it depends on how many species are present. Where there are several individuals but all of one species, they're fish. But when talking about more than one species, they're fishes.

that are of fish a meter or more long. In the Baltic Sea herring schools were reputedly so dense "that an axe thrust into their midst would remain upright."[3] Centuries of overfishing have since seriously depleted Europe's waters—the North Sea is estimated to support less than one-fifth of the small-bodied fishes there before fishing began, and populations of large species are reckoned to be down by a factor of 50. But as local fish stocks ran low, new technologies enabled fishermen to roam farther afield and discover yet-untapped riches.

When in 1497 the Genoese mariner Giovanni Caboto, better known to many of us as John Cabot, reached Newfoundland and became, as he thought, the first European to set eyes on North America, he found so many cod that his crew reportedly didn't need nets to catch them: they simply hauled them up in baskets slung over the sides of the ship. In fact, this Grand Banks cod-fest had probably been luring Europeans even before Cabot—it now seems that Basque fishermen had been crossing the Atlantic to those same fishing grounds every summer for a couple of centuries, but keeping their lucrative secret to themselves. Yet over time, even the exceptional fish stocks of the east Canadian and New England seaboards have since been all but destroyed.

Different methods for attempting to reconstruct the size of past populations from historical information suggest that on the continental shelf off Nova Scotia, cod stocks have fallen by 96 percent in the last 150 years, and farther north around Newfoundland, a mind-boggling 7 million tonnes of cod present at the time of Cabot have been reduced to just 200,000 tonnes—just three one-hundredths of sixteenth-century levels. The irony of this is that cod are phenomenally fecund animals—females can produce up to 9 million eggs annually and live for 30 years—so they can withstand a higher level of harvesting than most creatures. Yet in the face of unrelenting fishing pressure, East Coast cod stocks continued to nose-dive, and by 1992 the Canadian government had no choice but to close the Grand Banks fishery completely. Almost 2 decades later, there's no sign of any recovery. It seems that the riches of a place once known as *Bacalao*—the land of the codfish, which fed people from New Hampshire to Nigeria, from Jamaica to Portugal have gone forever.

Fishing has of course moved on again—to the last, vast frontiers of the mid-oceans. The *Andrea Gail*—the swordfishing vessel at the center of the

---

3. O. Magnus, *Historia de Gentibus Septentrionalibus* (*Description of the Northern Peoples*), vol. 3, translated by P. Fisher and H. Higgens (1555; London: Hakluyt Society, 1996), quoted in Roberts, *The Unnatural History of the Sea*, p. 21.

book and film *The Perfect Storm*—was so much at the mercy of the weather in large part because progressive depletion of its quarry had forced it to hunt way out in the Atlantic, six days from its home port. But even in these immense spaces industrial fishing has caused stocks to plummet. Across the entire Atlantic, mathematical reconstructions suggest that large predatory fishes like cod and halibut—the most valuable and therefore prime target of fisheries—have declined by nine-tenths since 1900, with two-thirds of that fall happening since 1950 alone. Detailed analyses of catch records from the mid-oceans support the models, and show that populations of species like tuna, swordfish, and marlin typically underwent 80 percent collapses within just 15 years of the arrival of intensive fishing operations.

With ever-more sophisticated technologies for finding fish and ever-more powerful gear for catching them, the industry has responded by fishing farther down as well as farther out—bringing new, deep-water species like orange roughy and monkfish into the market. And with declines in the large and palatable top predators, there's also been a progressive shift in interest to other parts of the food chain—so that now in addition to targeting species like salmon and tuna and smaller predators like herring and mackerel, we are increasingly turning our attention to clams, mussels, and even jellyfish. In addition, more of the previously discarded bycatch—the accidental take—is landed and sold.

To make these new, unusual, and sometimes downright unattractive species more palatable to consumers, there's been a rebranding campaign too. What's now sold as orange roughy was until recently known (less appealingly but more informatively) as slimehead, on account of the mucus-filled canals on its head. Likewise, spiny dogfish has become rock salmon, Patagonian toothfish has been reinvented as Chilean sea bass, and the small flatfish formerly known as witch now goes by the name Torbay sole.

Yet despite the increasing breadth of what we'll pay to eat, and despite fishermen traveling farther, fishing deeper, and generally trying harder than ever, it seems that time is running out for many marine fisheries. Even with all this increased effort, the global marine fish catch—after decades of growth—has at best leveled off, at around 80 million tonnes a year. But these are Food and Agriculture Organization data, which, as prominent marine scientists Reg Watson and Daniel Pauly at the University of British Columbia have pointed out, may be distorted by systematic overreporting by Chinese officials.[4] Factor that in, they say, and because China catches

---

4. Until the late 1990s Chinese officials were apparently promoted (or not) based on whether they achieved prespecified production increases. In fisheries they therefore had a strong incen-

far more fish than any other nation, the adjusted figures show a sustained decline in worldwide catches of between one- and two-thirds of a percent every year since the late 1980s.

If catches are dropping while effort is going up that means only one thing: fish stocks themselves are falling even faster. That's bad news not just for the target species themselves, but for people as well. When the great Grand Banks cod fishery collapsed, an estimated 40,000 people lost their jobs. Income support and retraining cost the Canadian taxpayer over US$3 billion. In many other places, fishing is more labor-intensive, so the importance of fish to human livelihoods is even greater. Developing countries make more money from seafood exports than from exports of coffee, cocoa, rubber, tea, meat, and rice combined. Worldwide, around 35 million people depend on fishing for their living.

And declines in major fisheries have wider knock-on effects too. The ripples reach far inland. Although demonstrating causality is hard, it's been argued that the early 1970s slump in Peru's vast anchoveta fishery in turn helped trigger the clearance of large tracts of Brazil for soy production by causing a rise in the price of soybeans—one of the few oil-rich substitutes for fishmeal in animal feed. More recently, industrial overfishing of West African waters[5] has been linked to a decline in local fishermen's catches, a rise in the price of fish and meat, and an unsustainable increase in hunting of so-called bushmeat—species like antelope, porcupine, and baboon.

Back at sea, other organisms pay a hefty price for overfishing as well. On average, for every 10 tonnes of the target species landed, another tonne of dead or dying bycatch is thrown overboard—nontarget fish, but also seabirds, marine mammals, turtles, seasnakes, and a multitude of invertebrates. And some fisheries are far worse than the average. Shrimp-trawling in the tropics, for instance, catches up to 10 tonnes of bycatch for each tonne of shrimp, much of it composed of young fish of valuable species that consequently don't live long enough to be caught by other fishermen.

Habitat damage from the immense fishing gear now used is even more worrisome. Consider the questionable practice known as bottom-trawling. It may sound comical, yet when it involves 10,000-horsepower ships towing heavily weighted nets the size of football pitches across the ocean

---

tive to overreport catches—and for 20 years Chinese catches did indeed show consistently (and, in global terms, anomalously) high growth.

5. A third of this industrial fishing is done by foreign vessels from the European Union and elsewhere, which, having depleted stocks nearer home, have benefited from taxpayer-subsidized agreements allowing them to fish in West African waters.

floor, bottom-trawling is anything but. Some areas are trawled four or more times each year—marine scientist Jeremy Jackson says that his colleagues at the Scripps Institution of Oceanography daren't leave instruments on the seabed for more than a couple of days for fear of them being swept up in a net. And the effects can be devastating. Bottom-trawling has been likened to clear-felling a rainforest—completely destroying the structural complexity of the seafloor, and the creatures (the sea fans, the sponges, the burrowing worms, the cold-water corals . . .) responsible for it. In places, entire deep-water coral reefs many thousands of years old have been razed and replaced simply by mud, lined with the deep tracks of the nets.

The Whys of Unwise Use

So why on earth is this happening—the serial exhaustion of fisheries, the resulting loss of people's livelihoods, and the associated obliteration of marine habitats? There are many contributing factors, but to my mind they come in three main flavors. First is the pervasive problem that surfaced when I talked about ecosystem services: the problem of externalities. Many fisheries—especially those on the high seas[6]—are classic examples of open-access resources, which are available to be exploited by anyone who can reach them. This means self-restraint simply doesn't make sense. If I refrain from taking a few extra fish today, before I can come back to enjoy the benefits of leaving behind a healthier stock someone else will have done so. The benefits of my good behavior—or the costs of me instead catching more fish right away—accrue to other people, not me. So selfish but rational decision making militates against sustainable harvesting, in favor of mining as if there is no tomorrow.

And exactly the same argument holds in a temporal sense too. Why should fishermen limit their catches now to ensure there are plenty of fish available for future generations, when the beneficiaries are unlikely to be their own relatives or friends (or increasingly, even their own countrymen)? The reality is that unless individuals or groups of people that harvest resources have secure and exclusive access to them, they're unlikely (except when forced by regulations) to manage them for long-term persistence.

---

6. The high seas are formally those waters beyond the jurisdiction of any country. They start 200 nautical miles out from each landmass—the limit of the relevant country's Exclusive Economic Zone.

Second, even if you wanted to exploit a fish stock sustainably, there are serious technical difficulties in identifying a safe level at which to set your offtake. Conventional theory suggests that you can get Maximum Sustainable Yield (MSY) when population sizes are substantially below their prefishing level. Even though this means there are fewer breeding adults, there's also less competition for food, so more offspring survive to adulthood and hence can be harvested sustainably.

But if fishery managers get their stock assessment wrong (easily done when the animals involved are spread over vast areas of the sea), they might recommend too high an offtake, which could drive the population below its MSY size. The benefit of yet lower competition for resources will then be outweighed by having far fewer adults left to breed, so the population becomes unable to replace the animals being harvested and therefore declines still more. Add to this tricky MSY balancing act the extra complications caused by unpredictable fluctuations in ocean temperatures, currents, and so on, plus the fact that real-world fish communities comprise many species (whereas most fisheries models deal with only one), and it becomes self-evident why striving for a theoretically maximum harvest is a hazardous enterprise.

The third problem is one of overcapacity. When excess fishing causes catches and profits to fall, one solution is to fish less—but this only helps the individual fisherman (and fish populations) if everyone plays the same game. Far more often, the response is to increase fishing effort. This keeps catches up, but only temporarily. Because everyone else will increase effort as well, this will drive stocks further down, and eventually catches will fall. The net effect: like Lewis Carroll's Red Queen, fishermen end up running ever harder simply to stay in the same place.

This inexorable ratcheting up of the power of vessels and the technology they deploy has been fueled not just by rational (albeit short-term) decision making by fishermen but by public subsidy too. In many countries, faced with the prospect of falling catches, politically powerful fishing interests have appealed to governments to help—which have often done just that. In total, subsidies to fishing (in the form of grants, favorable access to credit, subsidized fuel and so on) now run at over US$30 billion each year[7]—nearly all of it to large-scale industrialized fisheries. The end result? By some estimates the global fishing fleet is twice as big as it needs to be to land the

---

7. This, to land a global marine catch worth around US$90 billion a year at first point of sale. In total, across fishing, farming, forestry, water, energy and transport combined, perverse subsidies such as this—which though politically expedient are both environmentally and economically damaging—have been estimated to run to a staggering US$2 trillion a year.

current worldwide catch. There's more pressure on fewer fish than ever before—and taxpayers have unwittingly footed the bill.

Evidently, we need a sea change in the way we deal with the sea. Without it, prospects look bleak indeed. Based on the growing frequency with which fish populations are collapsing, one controversial study by a group of highly regarded marine scientists has projected that under business as usual just about every currently exploited stock will have collapsed by the middle of this century. Commercial fishing at sea—at least for fish, if not for the jellyfish and krill we will move on to when they're gone—may during our children's lifetimes become merely a memory.

Yet much can be done to improve the situation for marine biodiversity and fishermen alike. Eliminating unregulated access to fishing grounds and increasing fishermen's sense of ownership of the stocks they rely on (internalizing those externalities again . . .) can encourage more farsighted management. Tighter controls by governments and others on how much fish can be harvested, when, and where from, can help stocks rebuild. And though politically difficult, retiring superfluous boats and eliminating perverse subsidies (which help only today's fishermen, but harm everyone else) can shift the focus from the immediate horizon toward a long-term future for fishing communities.

At the annual meeting of a small fishermen's association on the West Coast of North America I find out about one other exciting and innovative approach—one that links right back to my disconcerting supermarket experience and that involves, of all things, a high seas fishery targeting tuna.

Hunting for a Silver Bullet

Scott Hawkins, skipper of the *Jody H.*, has a vice-like handshake that comes as something of a surprise given his gentle voice. He explains why fishing matters so much to him: "I'm a third generation tuna fishermen—my grandfather died at sea. So for me, it's in my blood. I started fishing albacore with my dad when I was eight." (A brief point of explanation: albacore is a medium-sized species of tuna found throughout the world's warmer oceans, and Scott and his fellow San Diego–based albacore catchers search for them across a continent-sized expanse of the Pacific Ocean. So these weren't exactly your ordinary father-and-son fishing expeditions.)

Scott carries on: "I did a couple of trips each summer after that— sometimes more if one of the crew got thrown in jail and my dad needed extra help. By the time I left high school, though, the fishery was beginning to decline, and I figured I wasn't going to be a fisherman. But then a friend talked me into one more season—and I stayed on for another six summers. I ended up buying my grandfather's boat and then, finally, I knew I was going to carry on being a fisherman." He looks down, smiles wistfully, and says, "Of course my wife knew already. She'd seen the look on my face when I was out there—what I looked like when I was catching fish, or even talking about fishing. As soon as I untie the boat and I can't see land any more I have an inner peace. Being at sea . . . it just makes me alive."

Peaceful isn't how I'd describe what I find out about albacore fishing, but then fishermen like Scott aren't ordinary people. For one thing, they do their job in some of the most remote corners of the earth. Like the other six commercially important tuna species, albacore are ultrafast, pack-forming predators that roam vast areas in search of their prey. This means those who hunt these super-streamlined nomads in turn usually do so way out into the ocean. In the case of the San Diego albacore boats, they travel as far as the international dateline, a staggering 6,000 kilometers from port.[8] To find the widely scattered shoals they look for places where the ocean appears to be boiling—where the tuna have panicked their quarry into bait balls pressed hard against the surface of the sea and are attacking them from all sides.

But searching for fast-moving needles in far-flung oceanic haystacks is where the similarities with most other tuna fisheries end. The bulk of the world's tuna catch—the skipjack and yellowfin that end up in hundreds of millions of cans each year—comes from industrial-scale purse-seining operations, where a circular net up to half a kilometer across and 200 meters deep is set around an entire school of fish, and then drawn tight shut. The resulting haul can contain as much as 100 tonnes of tuna (an astounding amount of fish, when you start to think about it). But of course it often includes thousands of unwanted bait-fish too, not to mention sharks and any other predators drawn to the feeding frenzy.[9] Other tuna fisheries—

8. Nowadays, though, Scott's colleagues tell me they spend much more of their time "only" 500 to 1,000 kilometers out from the Oregon and Washington coasts. "Near the beach," one of them calls it. Albacore humor is pretty dry.

9. Until the early 1990s purse-seiners in the eastern Pacific, mostly from Mexico, used dolphin pods to locate tuna, which often swim underneath them. Nets set around dolphins caught plenty of tuna but killed millions of spotted, spinner, and common dolphins too. Since then, demand for dolphin-friendly tuna has forced many fishermen to switch to deploying large man-made rafts called fish aggregation devices, under which (for reasons nobody yet understands) all sorts of fishes gather. Purse-seining around FADs catches lots of tuna and very few dolphins,

especially those after bluefin, the most highly prized (and consequently rarest) species—instead use longlines. Up to 100 kilometers in length and set with thousands of baited hooks, these again often cause severe collateral damage in the form of bycatch—turtles, seabirds, and many nontarget fishes. But neither purse-seining nor long-lining are much good at catching albacore, and so the West Coast albacore fishery is a very different (and far less damaging) operation.

"What we do is hands-on fishing," Scott's colleague Bobby Blocker tells me on a tour of his boat, tied up in San Diego Harbor. Bobby is white-haired, half-Mexican and half-German, and has spent nearly 40 years chasing tuna around the Pacific. He fishes for albacore between June and October, when 3- and 4-year-old fish migrate eastwards across the ocean, feeding and fattening up as they go. *Her Grace*, Bobby's boat, is all white and bristles with antennae and radar equipment. The smell of varnish and paint tells me she's undergoing some between-season maintenance. But the main thing I notice is her size. Bobby and his crew go out for a month or more at a stretch in seas I can scarcely imagine in a 30-year-old vessel just 20 meters long. And *Her Grace* is one of the bigger and newer boats in the fishery.

Bobby describes how they catch albacore in two different ways. Troll-and-jig fishing involves towing a few dozen short lines out the back of the moving boat. There's just one barbless hook per line, which is hauled up whenever a fish strikes. It sounds a bit like a mackerel fishing trip I once took off England's south coast, except the fish and the waves are both ten times bigger. And you don't get to go home for tea.

Bobby's team specializes in the other technique—pole-and-line fishing—in which three or four crew members stand in racks suspended over the ocean, each using a flexible pole equipped with a line and a barbless hook to catch the fish directly from the surface. "Fishing doesn't get any more basic than this," he enthuses, as we chat in the wheelhouse. There's a stack of high-tech equipment up here, but down in the racks it's different. "It's just one man, one hook, one fish." As soon as a fish is caught it's flicked overhead and onto the deck and the crewman casts his line again—landing, in a good spell, 5 or 10 fish every minute. Doing this for an hour or more, without hitting a fellow crew member with a pole or an 8-kilo flying tuna, demands exceptional dexterity and strength. Doing it over the side of a heavily pitching boat, with the rack rising up then sinking waist-deep in

---

but even more bycatch of other animals (turtles, sharks, swordfish, marlin, barracudas, manta rays . . .). Dolphin friendly maybe, but distinctly unfriendly to just about everything else in the sea.

water, when you can hold on only by bracing your feet under the guard rail, requires stomach muscles and a level of fearlessness I can only marvel at.[10]

There are other highly skilled tasks besides landing the fish. There's the chummer, whose job is to throw live bait-fish (a.k.a. chum) out on the surface at just the right rate to keep the albacore biting and focused on food. "He's loading the guns," Bobby explains, "so he has to be in time with the skipper, who's working the sonar and the meters." And of course the skipper has to find the albacore in the first place. "I spend a lot of time scanning for signs," he says, "changes in the water surface. When the fish are coming up it's like looking out at a lawn after it's been freshly cut." He cues in on wildlife too. "Every bird signifies something. I don't know the names of all of them, but I sure know what size bait they each feed on."

Having found an albacore shoal, the real skill is staying with them—feeding and catching them without panicking them into bolting. "If you scare them, they're like deer running from a fire—they won't stop to graze. But if you get it right," Bobby adds, "they bite like piranhas. When they switch over to feeding like that, they go blind. You pull 'em up 'til your arms ache. And by then . . . well, you hope the boat's full."

From the point of view of the oceans, the great thing about such a labor-intensive fishery is that it takes a limited and highly selective catch. When a purse-seine is set right, it swallows the entire shoal, leaving behind nothing but water. But with pole-and-line fishing or trolling the fish can only be caught for as long as they're feeding. As soon as they lose their appetite they dive, so plenty of fish escape. One boat owner tells me, "Out of a 30-tonne school we maybe only pull out a couple of tonnes." And bycatch, he explains, is exceptionally low. "Last year I caught 300,000 pounds of fish. Apart from two horse mackerel, everything was albacore." Another skipper tells me he hasn't caught anything but a tuna since 2001.

Yet despite its own limited impact, the West Coast albacore fishery has still met with major problems from industrial fishing. In the 1980s and early '90s, Japanese and other Asian vessels operating in the Pacific started using immense drift nets, sometimes tens of kilometers long, to catch squid. The problem was that these "walls of death," as they've been dubbed, tangle and kill everything in their path. This completely indiscriminate fishing method hit albacore stocks hard, reducing the average catches of the West Coast boats from around 100 to just 20 tonnes per season. Even those fish they did manage to catch were often scarred escapees from drift

---

10. To see and marvel for yourself, visit YouTube and type in "AAFA tuna fishing." But don't do so if your sea legs aren't so good and you've recently had lunch.

net encounters. The albacore fishermen and others petitioned their governments, with the result that since 1992, high seas drift netting has been banned by the United Nations. Although there are new fears it's being revived illegally, for now albacore catches seem to have recovered.

But industrial fishing—this time in the form of the yellowfin and skipjack purse-seining operation—has hurt the albacore fishermen in another way. Albacore get sold into the same supply chains as the other two species, so with the dramatic growth in purse-seining flooding the market with cheap tuna, and with fish buyers therefore in a position to haggle the price down right until the fish get unloaded in port, the price of albacore has fallen through the floor. In 1981 a tonne of West Coast albacore fetched US$2,300 at dockside. Twenty-five years later, it was down to just US$1,275.[11] Meanwhile, fuel prices trebled, so profits had all but disappeared. Skippers were leaving the fishery as fast as they could sell up, and generations of albacore fishing looked like it was coming to an end.

A group of West Coast fishermen decided something had to change. Two dozen vessel owners, including Scott and Bobby, got together and formed the American Albacore Fishing Association. They elected larger-than-life skipper Jack "Bandini" Webster as their president and his diminutive but extraordinarily energetic wife (and mother of five) Natalie as secretary. Even though AAFA members hardly see one another—they keep in touch by radio but fish over a huge span of ocean and meet up only at annual events like the one I've come to—it's immediately obvious they're a close-knit community. Over a beer in the conference bar, Webster explains. "These guys are my 9-1-1. If I get into trouble out there [I think he's referring to the mid-Pacific, not the bar] the only person that's going to save me is one of these guys. And I'm the only person that's going to save them." Then it dawns on me. I've seen this sense of mutual self-reliance before—among the inhabitants of Loma Alta. And just as there, strong group decision making is helping to reinforce the careful use of natural resources.

Natalie and her members realized that the only way out lay in the albacore fishermen making the most of what they did—which is catch high-quality tuna more sustainably than just about anyone else. They just needed to find enough consumers willing to pay them to do it that way. They'd already heard of a rapidly expanding global initiative called the Marine Stewardship Council. This attempts to reward sustainable fishing through an independent seal of approval. The AAFA fishermen hoped that if they could get the MSC to certify their fishery, that would give them pref-

11. Adjusting for inflation, that's just one-quarter of its 1981 price.

erential access to increasingly discriminating markets, which could in turn give them the two things they desperately needed: more money for their fish and, even more importantly, a guaranteed, stable price for the whole of a season's catch before they left port.

Applying for MSC certification turned out to be a smart move. Helped in part by a grant from the World Wildlife Fund, AAFA approached an accredited certifier, who in turn scored the fishery against MSC's three broad categories of sustainability: its impact on stocks of the target species (in this case, albacore), its impact on the wider ecosystem, and how well it was managed. AAFA passed with flying colors. As one skipper proudly tells me, "Our fishery was a no-brainer for the MSC." Even when it's successful, certification usually comes with some requirements for improvement. "But our only requirement was for more research. We were the first fishery to have so few conditions."

Meanwhile, Natalie worked on securing two other ingredients for success. She got AAFA audited to show that it was capable of tracking exactly where all of its fish ended up—all the way "from boat to throat"—so that it could qualify for another MSC endorsement, the Chain of Custody standard. This means AAFA albacore can carry MSC's blue-and-white logo (half fish, half tick of approval), and customers buying it can be confident it's not been mixed with fish from any other source.

And Natalie also went traveling—to the world's biggest seafood fair, in Brussels, where European buyers were keen to get hold of the first tuna ever to be certified by the MSC. Natalie tells me some of the long-term U.S.-based albacore buyers were worried by the competition and tried to make cut-price deals with individual fishermen. But they refused to break ranks and instead supported Natalie's efforts for fishery-wide sales agreements. It worked. Within two years of the start of the certification process, AAFA had favorable deals to sell its tuna in European markets—and Scott's albacore, beautifully packaged in glass jars and bearing the MSC tick,[12] began to appear on the shelves of the same local supermarket whose fish salesperson I'd harangued a few years before . . .

Fish and Ticks

The idea of the Marine Stewardship Council began in 1995 in, of all places, London's exclusive Groucho Club. Mike Sutton, an American conserva-

---

12. Alongside the inane but undeniable "Allergy advice: contains fish". . . .

tionist telling me this story in his office at the world-famous Monterey Bay Aquarium (a few hours' drive from the AAFA meeting), worked at the time for the World Wide Fund for Nature.[13] The man he met with was Simon Bryceson, an "issues management" consultant advising Anglo-Dutch giant Unilever. At that stage they supplied one-quarter of the entire U.S. and European fish market through well-known brands like Bird's Eye and Iglo but were beginning to wonder, in the wake of the Grand Banks cod collapse, how they could continue to put the fish in fish fingers. The men got together because they realized that their two organizations shared a common problem: how to enhance the sustainability of the world's fisheries.

After years working in public policy, Mike Sutton had come to the conclusion that governments alone couldn't save the oceans. "Fishermen are simply too powerful politically." Instead, he was persuaded by green business guru Paul Hawken's argument that the only force powerful enough to tackle unsustainable commerce is commerce itself. For its part Unilever had just lost £65 million in the UK's mad cow disease crisis, through the ensuing collapse in public confidence in beef products. Mike explains, "The last thing Unilever wanted was a similar problem with its seafood products. Moving toward more sustainable fisheries was simply good business sense. It was all about risk management and enlightened self-interest."

Mike and Simon discussed the model of the Forest Stewardship Council—a coalition of NGOs and forestry companies that had recently set about independently certifying and promoting sustainable logging operations. They wanted something more streamlined—in their view the FSC was getting bogged down in bureaucracy and procedure—but they could see that an alliance between their organizations just might work.

Mike recalls the case for a deal: "From WWF's point of view, partnering with Unilever would mean they could have an immediate impact on the fishing industry. For Unilever, working with WWF would ensure their efforts to become more sustainable had wider credibility. The two had different motivations, but that didn't matter, because they had the same goal: to provide powerful incentives for conservation of the oceans." Back at Unilever one board member reportedly said that trying to tackle the fishing crisis would cost the company millions. A second member agreed, but

13. Like unappealingly named fish, conservation organizations have a habit of changing names. In 1986 the World Wildlife Fund was renamed the World Wide Fund for Nature—except, confusingly, in the United States, where it still uses its original name. Thankfully, in 2002 another, rather different source of confusion was eliminated when, under legal pressure, the World Wrestling Federation stopped calling itself WWF and became instead World Wrestling Entertainment.

pointed out that doing nothing would end up costing a great deal more. Within a year both WWF and Unilever were persuaded, and in 1997, with £1 million start-up investment from each parent and a high-profile chairman in the form of John Gummer (recently retired from being the UK's longest-serving environment secretary) the Marine Stewardship Council was born.

Over the next two years the MSC went on the road, running workshops in eight countries to get input from as many experts as possible on the design of the "MSC Standard"—the framework for deciding whether a fishery is sustainable and so merits certification. The resulting document set out what MSC terms its three principles (the three tests that AAFA had to pass): the impact of the fishery on the target stock (principle 1); its impact on the wider ecosystem (principle 2); and the effectiveness of its management structures (principle 3).

Under this system, to become certified, a fishery usually first asks an independent, accredited certifying company for a confidential preassessment, aimed at filtering out hopeless cases and pinpointing key areas for improvement. If the fishery decides to go into full assessment, the certifier's experts then score its performance against a set of indicators. Fisheries that get a high enough average score but only a low pass on one or more individual indicators get approved, but have to undertake follow-up remedial actions. All successful fisheries are inspected annually and need to apply for recertification every five years. For the products of certified fisheries to carry the MSC logo they also have to undergo—as AAFA did—a separate Chain of Custody certification to confirm their fish never get mixed with products from noncertified sources. Making both elements of certification the responsibility of third parties, making the standard transparent and ensuring the assessment process and results are open to objections from NGOs and other interested groups, are seen as central to the MSC scheme's credibility.

By 1999, with its standard in place and with growing interest from fisheries, fish buyers, and retailers, the MSC became independent of WWF and Unilever, and a year later certified its first fishery—for rock lobsters off the coast of Western Australia. Other certifications and more blue-and-white MSC ticks followed, for a series of small, traditional fisheries around the UK, but for much bigger fisheries as well—for wild salmon in Alaska, for hoki around New Zealand, and for Patagonian toothfish in the waters around South Georgia in the South Atlantic Ocean. There were increasing levels of commitment from the other end of the supply chain too. As early as 1996 Unilever had promised that by 2005 it would be sourcing all its seafood from sustainable fisheries. Other major fish processors (such as Young's Seafood) and several major retailers (including Whole Foods Market in the United States, Sainsbury's and Marks and Spencer in the UK, Edeka and Lidl in Germany, and Coop in Switzerland) began to make similar commitments.

Yet progress with certification was slow, raising questions about whether retailers' ambitious targets could be met (and if supply couldn't keep up with demand, whether demand might fall away). The MSC itself nearly went bankrupt, and conservationists and the fishing industry voiced serious (though predictably rather different) concerns about its effectiveness. NGOs complained about specific certified fisheries, saying they didn't merit MSC approval. Meanwhile industry voices argued the certification process was overly cumbersome and expensive. Two of the MSC's major funders commissioned independent reviews, which made strong recommendations for greater consistency in applying the MSC Standard, tougher enforcement of remedial actions, and better data on whether certification was really making fisheries improve their practices. The MSC responded in striking fashion, and in 2004 changed its chairman, replaced its director, and initiated a wide-ranging overhaul of its governance and its certification procedures. The result was a dramatic step-change across the entire organization.

Commitment Issues

The man universally credited with reinvigorating the MSC is Rupert Howes, a fortysomething father of four and, in Mike Sutton's words, "the most energetic CEO I've ever seen." I meet up with him at MSC headquarters in an unprepossessing building opposite London's Victoria coach station. For an outfit aiming to save two-thirds of the planet, it's clearly not wasting any money on fancy offices. Rupert is immediately engaging—humorous, charming, interested in my project and brimming with enthusiasm about his own. "Of course we need other changes in fisheries—better management, properly enforced policies, and the eradication of illegal and unregulated fishing. But we live in a capitalist society, so a market-based approach has to be part of the solution too."

Since taking over, he has responded to critics by guiding the organization though a major review of its certification process. Though infectiously positive, he admits dealing with criticism from many quarters at once is difficult. "We're trying to steer a middle course here, and it's a bloody tough space to operate. But then again," he reflects, "transformation is a tough business. And what keeps me working 70 hours a week is the firm belief that we are making a difference."

During the first couple of years under Howes's leadership the number of certified fisheries doubled—and with increasing availability of products bearing the tick of MSC endorsement, interest from fish sellers grew too. The watershed came in February 2006, when Walmart, the world's largest retailer, declared that it was aiming to sell only MSC-certified seafood within five years. Mike Sutton explains, "The announcement sent shockwaves through the seafood industry." By itself Walmart doesn't sell that many fish, but the operations that supply it—and that depend on Walmart's custom—include most of the major seafood buyers. "At the time of the announcement many of these companies either had no knowledge of the MSC or didn't deal in certified seafood." According to Howes, the Walmart deal changed all that. "Because it was time-bound and ambitious, suppliers had to take the deal seriously. It focused minds."

Retail, however, is only half the story: in the United States, over 50 percent of all seafood is eaten in restaurants, so the catering sector is vital too. But then, just weeks after the Walmart breakthrough, the Compass Group—the largest catering company in the Americas, which supplies thousands of restaurants, schools, and hospitals—made a similar announcement. Sustainable seafood had entered the mainstream.

Yet while this remarkable scaling up of the MSC enterprise is impressive, I'm just as interested in understanding what's driven different elements of the industry to get involved. For the fishermen themselves, the perceived benefits are fairly obvious. As I found with the albacore hunters, it's partly the hope of an increase in price through responsible retailers and conscientious consumers paying a little extra. Far more, though, it's about security: price stability, continued access to increasingly discriminating markets, and, in cases where regulations are becoming tighter, valuable external evidence that a fishery should still be allowed to operate.

Realizing these benefits depends on the commitment of the retail and catering sectors to switch toward buying only from certified fisheries. Not everyone I speak to is convinced that the big-business pledges are entirely genuine. "Greenwash" is a term I hear several skeptics use. They talk of companies seeking short-term PR gains with no commitment to fundamental change. But Scott Burns, a burly Chesapeake Bay crab fisherman-turned-lawyer, and one of the figures behind the Walmart deal, disagrees. "They're still totally committed," he says, over breakfast at the Double-T Diner in Annapolis. "They're not backing away from the deal." I press him, asking if the retail shift toward sustainable seafood could be just a fad. He smiles. "It's been around for a while if it's a fad. And over that time more and more businesses have become engaged. I can understand the skepticism, but this thing isn't going away." So what is it that's really driving companies like Walmart, Sainsbury's, and Compass?

Strikingly, one thing that isn't very important so far, Scott tells me, is consumer pressure. One supermarket buyer I ask says this is starting to change, but it will take a long time. Mike Sutton agrees, and says that the real demand comes from the companies themselves. "This movement is happening because it appeals to enlightened businesses, who are then triggering a domino effect." There are sound commercial reasons for farsighted companies to back initiatives like the MSC. At its most basic, they want to sell fish, and they realize that to still do so in 10 or 20 years' time they need to switch now to buying from sustainable sources. There are also marketing benefits, from businesses being able to portray the seafood they sell as good for the environment. And because animals caught in smaller quantities using less destructive methods often arrive at shops or restaurants in better shape, there can be gains in terms of fish quality as well.

Scott Burns believes there are nonfinancial considerations at play too. "Altruism is important. Take the CEO of Walmart: he's an outdoors person, he has a personal interest. There are limits, of course—this is business— but the key is getting the leaders engaged. These are smart people. They're interested in what's going on in the world, and they get it." Being environmentally responsible can also pay off elsewhere in the workforce, Scott says. "You have to remember that out of their offices these are ordinary people. They care. So being sustainable can help companies attract and keep good staff."

The role of caring employees, of enlightened leadership, of maintaining a license to operate, and of taking a long-term view of profits—these are the same as the answers I got in Australia when I asked why Alcoa is striving to be the greenest mining corporation. And there's one other driver that the two stories have in common: the power of pressure groups. In Alcoa's case, concern about environmental backlash strengthened the company's resolve to continue to improve its act. In the case of sustainable fishing, one particular incident forced many companies to think about an issue they'd until then largely ignored.

Late in 2005 Greenpeace produced a league table ranking the largest UK supermarkets on the sustainability of their seafood. High-end chains Marks and Spencer and Waitrose came top, Sainsbury's scraped a pass, and Greenpeace graded all others as failing. The following January Greenpeace protestors scaled the roof of bottom-ranking Asda's headquarters in Leeds, unfurled a banner asking "Will stocks last?" and won international publicity for the issue of fisheries collapse. As the chief buyer for Waitrose tells me, "That action changed the face of retailing. Up until then the entire industry had been labeled as bad. But suddenly they pitched supermarket against supermarket." The results were almost immediate. Asda swiftly agreed to remove several unsustainably fished species from sale. Within a year, when Greenpeace published its next league table, Asda had risen to fifth in the rankings. All other supermarkets had improved their scores too. Greenpeace has not felt the need to publish its UK league table again— though it is now targeting the slower-moving U.S. market in the same way.

The story echoes a familiar theme on my travels. As with fishing and mining (or regrowing forest in Costa Rica, or tolerating woodpeckers in the Carolinas), it seems we usually need sticks and carrots, together. Strident NGOs to strengthen the resolve of businesses and politicians, as well as people working with the system. As Scott Burns puts it, "The good cop, bad cop approach works. We need that diversity of voices in this movement."

Net Gains?

So is the MSC working? Perhaps unsurprisingly, that depends on who you ask. One fisheries expert who helps assess applications for certification tells me that fisheries in general think the MSC Standard is too demanding, while conservation organizations worry it's too lax. An MSC insider I talk with goes further, explaining that the organization is trying to satisfy four very different interest groups—fisheries, environmental NGOs, retailers, and consumers. Everyone wants things to move in their own direction. The trick for MSC is to stay somewhere in the middle. Inevitably that makes for tensions, but despite the four-way tightrope act there is real progress.

There's clear evidence that some fishermen are benefiting from certification. A couple of years after AAFA got the MSC thumbs-up and started offering the only certified tuna in the world, Natalie was able to negotiate a 30 percent premium for her members' albacore, compared with the typical dockside price. More importantly, the price was fixed in advance for the whole season. As one albacore skipper tells me, "This is a real glimmer of hope. We're taking control of our fish, rather than being at the mercy of the fish buyers."

Elsewhere, the external recognition that MSC certification brought to rock lobster fishermen in Baja California, Mexico, prompted the government to invest tens of millions of dollars in electricity, water, and road infrastructure for isolated fishing villages. AAFA-esque increases in the price paid for certified fish have been reported for a handful of fisheries. In one detailed study supermarket shoppers in London paid 14 percent more for their pollock if it carried the MSC logo. But price gains are by no means guaranteed. Improved market access, on the other hand, seems more widespread, with MSC approval opening up new markets, for instance, for Alaskan cod and salmon and German saithe, and helping persuade the European Union to lower the tariff it levies on imports of certified rock lobsters from Western Australia.

Improvements beneath the waves have been harder to track. MSC says this is because they mostly happen during the confidential preassessment phase of certification, for which data are not publicly available. However, a major assessment of the 10 longest-approved fisheries uncovered moderate numbers of gains even after certification. For instance, accidental killing of fur seals by New Zealand hoki fishermen has fallen. Bycatch of seabirds in South Africa's hake trawl fishery dropped by over half after certification required ships to employ bird-scaring devices and avoid processing fish and discarding the offal while nets were being set. And across the board the MSC is now trying to get better baseline data against which it can measure the impact of certification. They believe that by looking back at what fisheries were doing right at the start of certification they'll find evidence of greater gains.

But there are many criticisms too. From the fisheries perspective, certification can require a lot of data, time, and money. Assessment costs can run into six figures, and in the case of Alaskan pollock—admittedly the largest fishery in the MSC program—certification took more than 4 years. Having had only 4 accredited certifying bodies probably hasn't helped, although with 10 more approved recently, Rupert Howes tells me that average assessment times have come down to 18 months.

The high costs and data needs of certification raise another less tractable issue: it's tough for small-scale fisheries in developing countries and elsewhere to access the scheme. All told, these land about the same amount of fish as the big commercial operations, employ far more people, and are considered by many to be more sustainable—yet they rarely have the scientific information or money needed to demonstrate this through certification. As a result, while some small fisheries like AAFA have been certified (often with the help of grants), so far only a handful of these are from the developing world. The MSC response has been a three-year project culminating in the launch of what it calls its risk-based framework. This takes a precautionary approach to assessing whether a fishery meets the standard when available data are patchy. Pilot studies suggest it works, and several small fisheries—in Latin America, Europe, and even West Africa—are now being assessed for certification using the new tool.

Questions have also continued to be asked about specific fisheries. For example, after certification, one of two stocks of New Zealand hoki declined sharply. Many argued that the fishery should therefore have failed when it came up for reassessment five years later. Controversially, it passed, but with stringent conditions on managing the catch to allow for population recovery. This seems to have paid off—within two years the stock had rebuilt, with MSC supporters arguing this was partly because of the conditions attached to recertification. Recent recertification of the vast Alaskan pollock fishery has proved similarly divisive, with some environmentalists attributing local stock declines to overfishing, rather than being caused (as the fishermen claim) by the fish shifting their range out of the area. Certifiers approved the fishery for a second time, because they felt it was reducing its catch levels in line with the decline. Time will tell if this was an appropriate decision, though promisingly, stock levels do seem to be on the rise.

To its greener critics, these controversies illustrate broader problems with the MSC process. Keeping the certifying bodies independent of MSC is crucial to the scheme's credibility. Yet some I talk with suggest that certifiers have had too much freedom to select their own indicators of a fishery's performance and so have been able to engineer passes. MSC has responded by tightening its assessment procedures, so that they're now far more prescriptive, and specify an identical set of indicators for all fisheries. As Rupert Howes says, "We've made huge efforts to fix the bar, so it's less open to individual certifiers' interpretations."

Yet the critics go further and point out that very few fisheries have ever failed full assessment. For its part, the MSC counters that this is because unsustainable fisheries fail at the preassessment stage. Because of commercial sensitivities (fisheries might understandably not want others to know if they fail), these screening results are confidential. But when I probe, MSC officers tell me that around two-fifths of applicants are told at preassessment that they would fail the certification process. A further 20 percent are advised to make substantial improvements before entering full assessment (improvements that are confidential, and as such are not systematically recorded or credited to the MSC process). Just 40 percent of applicants are recommended to go straight into certification. So in fact the failure rate is sizeable.

From what I can see the MSC takes each of these concerns—about high costs, limited access for smaller fisheries, and potential inconsistencies in assessment—very seriously. They must, because they threaten to under-mine the integrity of everything they do. There's one last concern, however, where they accept that they may simply have to agree to disagree with their critics. This time the question is not about making sure the bar is properly fixed, but about how high it should be set in the first place.

Very few dispute the sustainability of many of the small fisheries that the MSC approves—they use low impact gear and can demonstrate they've exploited their stocks carefully for decades. Some vocal critics argue that the MSC should focus entirely on such small-scale, already-sustainable op-erations. They should never approve industrial-scale fisheries and should probably exclude any form of trawling as well. That would make for an easy life, but the problem is that only approving small operations that scarcely need to change isn't exactly going to make a major dent in the sustainability of global fishing practices.

Instead, the MSC argues, it also needs to work with less-than-perfect fisheries: finding the better operations, encouraging them to improve prior to full assessment and then, providing they score highly enough, stipulat-ing further improvements as conditions of certification. Engaging with the imperfect rather than simply endorsing the already-virtuous. Rupert cites an appropriately maritime adage: "A rising tide lifts all boats." I'm not sure about every single boat—I suspect quite a few would prefer to remain stub-bornly anchored to the bottom—but as a general philosophy my impres-sion is that it is moving things in the right direction.

Sea Change, or Drop in the Ocean?

So will MSC certification live up to its promise? The answer depends on two issues. First, on how far the MSC can maintain its momentum in getting better-performing fisheries certified, thereby enabling major buyers like Walmart and Compass to meet their targets, and catalyzing a virtuous spi-ral of expanding demand and increasing supply. And second—in the face of all those caveats and concerns—on whether widespread certification is actually capable of delivering substantial benefits for the world's oceans.

On the question of the realism of the MSC's ambitions and those it has encouraged industry leaders to adopt, it is significant that shortly before selling off its fish-selling business, Unilever, one of the MSC's founders, ad-mitted that it had failed to meet its target of selling only sustainably sourced seafood by 2005. It got over halfway there—a remarkable achievement in itself—and was honest about its failure. But it nevertheless missed its own

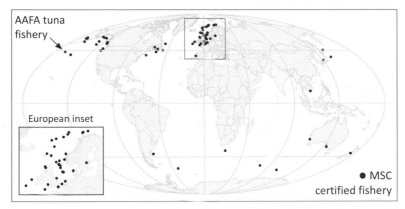

Map 8. More than 130 fisheries have been certified as sustainable by the Marine Stewardship Council, with as many again now undergoing assessment. Data from the MSC website (http://www.msc.org). Cartography by Ruth Swetnam.

target. And fish sellers' commitments have continued growing. The day before I visited him, Mike Sutton concluded a deal between the Monterey Bay Aquarium and Aramark, the world's second biggest catering company, for it to buy entirely sustainable seafood by 2018. Carrefour, the planet's second-largest retailer, has promised to increase the number of MSC-labeled products it sells. And celebrity chefs Jamie Oliver and Raymond Blanc have committed their flagship restaurants to becoming MSC certified.

Alongside this rise in demand, certification is also surging ahead extraordinarily quickly. In the first five years after Rupert became CEO the number of MSC-approved fisheries leapt from 10 to over 40. Over the next 2 years another 90 were added. Over 130 more are undergoing full assessment with, for example, the Danish Fishermen's Association declaring that all Danish fisheries will be MSC certified by 2012, and the Dutch government plowing more than €1.5 million into helping its fisheries through certification. The number of products bearing the MSC tick has grown even more steeply, more or less doubling every year—from 500 after 7 years to over 1,000 just a year later, nearly 4,000 after 10 years, and 10,000 products (even including MSC-approved cat food!) available in over 60 countries 12 years down the line.

Even in Asia, which consumes two-thirds of the world's seafood but where the MSC has been noticeably slow to get off the mark, demand and supply are now both growing fast. Four fisheries have been certified, more are in the pipeline, and in Japan, significant supermarket commitments have encouraged the number of MSC-certified products on sale to rocket from zero to 180 in under 2 years.

Globally, certified fisheries now catch more than 5 million tonnes of seafood annually, equivalent to almost 6 percent of all wild-caught fish caught for human consumption. Rupert Howes believes this will soon rise to 10 percent, and has his sights set on 20 percent. "That would get us to the point where there would be a large and diverse enough supply of certified fish to create a self-sustaining balance of supply push and demand pull."

Some industry commitments will be met. The main buyer for UK supermarket Waitrose tells me they're already achieving their target of selling seafood only from well-managed, low-impact fisheries, and believes this in part explains their impressive growth in market share. Whether Sainsbury's meets its commitment on time will probably come down to whether the large Maldives pole-and-troll tuna fishery becomes certified. Walmart's progress toward fully certified fisheries has slowed, and it's had to shift its target date back to 2012. Yet even when specific commitments are not met on time, they can have important signal value. Mike Sutton says, "This thing is like an oil tanker. We won't turn it on a dime—but serious commitments by leading companies bring others along too." Like nearly all the MSC champions I speak with, he's confident that certification will continue to expand, and that substantial industry commitments are here to stay.

But even with widespread certification in place will it be enough to make a difference? Some argue that the certification process itself is too imperfect and too vulnerable to being manipulated by a cynical industry that at its core is unwilling or unable to accept fundamental change. So certification can at best have only a negligible effect on what happens where it matters—in the water. Others—like marine conservationist Carl Safina—maintain that while the MSC is not perfect, it has made a real difference. "Without their efforts," he tells me, "nothing would be changing. It would still be like it was in the mid-1990s." For what it's worth, my own view from talking with many critics and supporters is that while caution is justified, there's growing evidence that the MSC initiative is succeeding in altering the way some fisheries and key industry players operate.

It's clear of course that certification can only be part of the solution, and that many other sorts of interventions will be needed to resolve the global fisheries crisis. Top of the list must be tackling the absurd overcapacity of the worldwide fleet—and so facing the politically uncomfortable truth of having to retire some state-of-the-art boats permanently (rather than simply paying them to fish elsewhere) and cut jobs now so that fishing has a future. Lowering or, better still, eliminating subsidies for overfishing and replacing them with incentives to harvest stocks sustainably (and as in the

Netherlands, helping such fisheries to get themselves certified) must be part of the answer as well. So too is taking much tougher action against illegal and unregulated fishing (which currently accounts for 1 in every 5 fish landed). Reducing permitted catches by lowering quotas, extending closed seasons, and designating many more no-take marine reserves (where fishing is banned and stocks can build up) is also important, as is involving fishermen directly in the management of their fisheries, and increasing each fishermen's stake in the future of their fishery—for instance by allocating so-called catch shares, which guarantee individuals a fixed proportion of the catch and so provide them with a direct incentive for letting stocks regrow.

But as with certification, the relative merits, practicality, and performance of most of these other courses of action are hotly debated. Some dispute the efficacy of marine protected areas. Others argue a system of catch shares can end up excluding poor or already marginalized fishers. It seems likely that different combinations of interventions are needed in different circumstances. But from what I've seen, certification certainly deserves a place on the menu.

The dramatic fall in accidental bird killings by South African hake trawlers and the recovery of New Zealand hoki seem in part the result of the MSC carrot. Certification has also led to new areas being placed off limits to Dutch plaice fishermen and the Isle of Man queen scallop fleet. The program obviously needs to get better at tracking its impacts (another consistent theme of many of the stories in this book), but as it does so it's likely that additional environmental gains will come to light. And in human terms the MSC scheme is certainly making a real difference to the lives of fishermen.

Onboard *Her Grace*, Bobby Blocker sums up what it means to him: "People are realizing that in a sustainable fishery you can fish for ever. Since MSC, the dominoes have started going our way. All of a sudden this boat has a value—it has a future." Scott Hawkins is an even more passionate advocate: "What MSC has given me is hope that there are people out here who appreciate what we do. I'm catching fish in exactly the same way as my grandfather did. And now, if my children get the sea bug, if they want to fish, then MSC can help them do so."

The key message I take from all of this is not to let the perfect become the enemy of the good. Conservationists are often one another's most vocal critics. While this can be constructive, it strikes me that where it undermines wider support for activities that, though imperfect, make the world a better place it can do more harm than good. The MSC approach is evidently not flawless—but as Rupert Howes reasonably points out, you can't expect perfection after just a dozen years. And the MSC is open about its problems and is working hard to fix them. It's a project with real and growing momentum—with fishermen, people who sell fish, and, increasingly, people like you and me, who simply want to feel good rather than guilty about what we eat. Because of that, unlike most of the other initiatives in this book, it's one to which almost all of us can make a direct contribution, every time we go shopping. And that's a good thing.

THE GLASS HALF FULL

So after all my travels, visiting six continents, and talking with dozens of people on conservation's front line, what have I learned? What insights have I gleaned from meeting fishermen and foresters, lawyers and scientists, peasant farmers and wealthy plantation owners? What did a gun-carrying ranger, a card-carrying journalist, an ex-poacher, and an ex-minister teach me that I didn't already know?

I started my journey with three questions. Are the good news stories I'd heard about genuine successes? What can they tell us about what works in a way that can help future efforts? And what collectively do they say about the prospects for wild nature: can it be saved, or is conservation doomed to failure? It's time for some answers.

Wins of Change

First, I've learned that conservation success is certainly possible— if rarely straightforward. Each of the places and programs I visited lived up to its billing. Not perfect, unequivocal wins with no wrinkles, rough edges, or lingering knotty problems. But I didn't expect that. The world is too complex, too fast-changing for absolute victories. Instead my examples are realistic, real-world successes. Some, doubtless, will yet be compromised. One or two may in due course be entirely reversed. But to date each has made great progress in stemming the tide of loss, and in each case nature is the better for them.

Of course the projects I picked are far from representative. Most other conservation efforts are less effective in slowing the threats that are unraveling the living world; globally, the rate of loss is, if anything, accelerating. On the other hand the circumstances of the examples I've explored are so varied that success clearly isn't confined to one particular set of favorable conditions or tractable problems. Indeed, I chose several of the stories specifically because they appeared to be making headway despite unpromising situations—giving nature more space in an already crowded country; making conservation on a massive scale a top priority for a new government elected instead to tackle crushing poverty and inequality; and curbing lucrative poaching in a desperately poor land plagued by stop-start armed insurrection. In short, winning against the odds.

The places I visited also illustrate how conservation itself is changing. In the past, it generally aimed to preserve nature for its own sake—to ensure the world is still home to rhinos and bitterns and the fantastically diverse flowers of the fynbos. That argument still has great traction and merit today. But increasingly there is a wider range of motives: from securing sustained harvests of wild-caught fish, through to safeguarding and restoring healthy ecosystems because of other, less tangible services they provide (such as more water when it's dry, less risk of flooding when it's wet, or space where busy people can simply unwind).

The range of methods and protagonists has broadened too. Not long ago conservation was largely the domain of scientists, lawmakers and rangers, who picked, designated and defended strictly protected areas. But greater understanding of the need for conservation outside reserves, and growing appreciation of its benefits, have expanded the cast of players and prompted new tools based on positive incentives. Being able to sell products to increasingly picky customers, getting paid directly for conserving creatures or ecosystem services, having less red tape in the way of doing the right thing, and so on—it's a long list. As a result conservation is now the concern of local communities, farmers, miners, engineers, and consumers. Crucially, it's becoming the business of corporations and companies too. In the jargon, conservation is gradually getting mainstreamed.

That widening outlook underscores one other important lesson I've learned from my journey: more than ever, in conservation one size does not fit all. So while a bottom-up, community-driven approach has obviously been pivotal to progress in the *comunas* of western Ecuador, in Assam it seems highly unlikely that rhinos would still be wandering its watery grasslands were it not for strong-arm, top-down enforcement of the law. Carrots are now rightly seen as vital, but sometimes sticks are necessary too. Likewise, while government has been an essential driving force for change in the Netherlands, Costa Rica, and South Africa, success in Alcoa's mines and among responsible fish retailers has been achieved almost entirely through a combination of private enterprise and NGO pressure.

THE GLASS HALF FULL   187

From what I've seen, those who advocate, on ideological grounds, this or that singular approach to conservation risk narrowing the opportunities for it to succeed.

Ways to Win Battles

So given that the stories I've looked at vary widely—in context, in threats, and in solutions—what if anything do they have in common? What insights, collectively, do the places I've been and the people I've met offer about how to improve conservation efforts, so that in future there might be more wins and fewer losses?

Unsurprisingly, the lineup includes several of the usual suspects. So the stories confirm that we often need a good crisis (rivers bursting their banks, rhinos failing to appear before visiting dignitaries . . .) to get things moving—although other examples (Alcoa's PR forecasts or models of the looming impact of alien plants) also reveal the power of the timely preemptive strike. The AAFA fishermen and the villagers of Loma Alta bear out social scientists' tenet that conservation is more likely to work where communities are strong and have fair and transparent ways of reaching decisions. And the vital role of Assam's wildlife laws, of the United States' Endangered Species Act, and of South Africa's Water Act all reaffirm the importance of having a strong regulatory and legislative framework—itself often the product of informed individuals and pro-conservation lobby groups pressuring governments to act.

These are all familiar faces from assessments of what makes conservation work. But seven other threads that have run through my journey strike me as worthy of more detailed dissection. Here they are.

1. **Great leadership.** All the cases I examined succeeded because of remarkable people. Exceptional men and women who imagined a better way, worked out how to get there, built relationships with the relevant people, and inspired others to follow their lead. Farsighted and passionate individuals like George White in Australia, Norman Moore in Dorset, and Frans Vera in the Netherlands, who foresaw a different world where concerns for nature and calls for its conservation would grow. Independent thinkers such as Richard Cowling, Dusti Becker, and Michael Bean, who could view the world through the eyes of post-apartheid politicians, Ecuadorian peasants, or Carolinan landowners and so see their way around seemingly intransigent problems. Early adopters and respected leaders of their com-

munities like village elder Don Alejandro, rancher Connie Jess, albacore skipper Bobby Blocker, longleaf grower Julian Johnson, and the board members of Alcoa and Unilever and Walmart. Where they led, others have followed, and attitudes have changed. And energetic, inspiring doers like Working for Water's Guy Preston, Rupert Howes at the MSC, AAFA champion Natalie Webster, and rhino defender Shri Lahan, who took on daunting tasks and made things work.

2. **Wherewithal and time.** Winning battles depends not just on leaders, but on patience and resources. The good-news stories I learned about have all involved taking a long-term view, recognizing that success is rarely a matter of months or even years but instead involves decades of hard work. Completely clearing South Africa's invasive plants will probably take the best part of a century; armed defense of Kaziranga's rhinos may be needed indefinitely. Doggedness is crucial too—several projects I looked at flourished but then ran into new problems that required fresh ideas and energy. And in nearly every case progress has hinged on adequate funding. Regrowing complex forests, battling Hydra-esque aliens, and reinstating nature across a crowded country are things that don't come cheap. Spend too little and protection gets weaker, landowners don't sign up to schemes, staff become frustrated. In India, a lack of money to upgrade "casual" guards to permanent positions led to Golap Patgiri becoming a poacher. More generally, studies of how well parks protect the species they've been set up to conserve consistently identify money as a limiting factor. Everyone who's examined the figures has concluded that in most of the world conservation is woefully underfunded. If we're seriously concerned about saving the planet we need to go much further in putting our money where our angst is.

3. **Being bold.** Many of the successes I visited worked in part because they dared to think big. Their proponents dreamed of a nature network spanning a whole country, of a restored forest every bit as rich as what went before, of an alien clearance program employing tens of thousands of people, or of a certification scheme that might extend to one-fifth of all the world's wild-caught fish. With ambition comes the risk of failure—or at least of having to compromise to meet targets. But many argue that as the scale of the threats to nature grows, so too must our responses. We must raise our sights. Once conservation projects involve tens of millions of dollars, people who make the really big decisions—in agriculture departments, banks, transnational corporations, and finance ministries—become interested;

nature moves up the agenda. Conservation organizations are realizing this: mega-projects are becoming an increasingly common part of the conservation landscape.

4. **Relevant research.** Properly knowing about a problem is also important, at several stages. It usually takes scientific data to document change and establish whether it's a real problem or simply some natural fluctuation. Careful research is then typically the best way of identifying the most promising actions for dealing with threats and reversing losses. In some cases—as with the fynbos ecologists' water calculations, the Arnhem Land fire results of Peter Cooke and co., and Dusti Becker and her fog interceptors—smart science can uncover additional, game-changing arguments for conservation action. And persistent monitoring matters too, for detecting change in the first place, discovering whether interventions are working and suggesting refinements, and providing supporters with evidence that their backing is yielding dividends. Worryingly, some of the biggest initiatives I looked at appear to spend very little on such long-term tracking. But funding everywhere is becoming increasingly linked to evidence of results, so this seems like a terribly ill-advised saving. That said, while having enough research is important, there can also be too much of it. Sometimes academic scientists are reluctant to commit to practical recommendations without more research, and governments are often all too happy to pay them to measure and model while Rome burns if that delays unpalatable action. Often it's best to do a bit of science, and then take action and monitor to see if it's working. Not too much, not too little: a Goldilocks-sized portion of research.

5. **A problem shared . . .** For me the most important and interesting theme common to almost all the successes I've sampled has been the ability of their architects to think creatively. Rather than focusing solely on the needs of threatened creatures and places, they've looked at problems from the perspective of the people involved—local communities, politicians, landowners and business leaders. Through understanding those individuals' concerns and constraints, innovative conservationists have been able to devise novel solutions that meet several needs at once and so broaden the buy-in for conservation. In South Africa the stroke of genius was to recognize alien trees as water wasters threatening the nation's economic growth. That transformed their eradication from an exercise in conserving plants into an opportunity to invest in poverty reduction on an unprecedented scale. In Arnhem Land the answer came from realizing that restoring tra-

ditional fire management wasn't just beneficial for sensitive vegetation but could simultaneously lower greenhouse gas emissions and help an aboriginal community back onto its land. Safe Harbor happened because people took the trouble to take landowners' worries about their liabilities seriously. And the MSC came about because conservationists treated as sincere Unilever's concerns about the sustainability of its supplies. Conservation needs all the friends it can get, and broad, imaginative thinking seems a powerful way of finding them.

6. Playing politics. Conservation is usually about changing the way things are done, and so it almost inevitably involves politics of one sort or another. Several of the projects I looked at were helped by their authors' tactical astuteness. In the early days of Oostvaardersplassen, Frans Vera and his fellow Dutch musketeers worked out how to play the system, exchanging information and keeping one step ahead of their political masters. In South Africa, the scientists who calculated the scale and urgency of the problem posed by alien invaders had the wit to nevertheless wait until the arrival of a new government, receptive to radical ideas, before launching their bid for a quantum leap in clearance efforts. In my experience most conservationists are gentle and earnest people. That's a great thing, but it seems being politically savvy and having a shrewd sense of timing can be important assets too.

7. Improvement, not perfection. The last lesson I take from my travels is the importance of not letting the perfect become the enemy of the good. All the efforts I profiled here have their flaws—that's inevitable. Whatever we'd ideally like to do, practicalities and politics intervene. Costa Rica's PSA scheme is generally agreed to be too broad-brush and hasn't yet managed to target those forests that are more likely to be converted; the MSC would benefit from better understanding how far (and under what circumstances) certification makes a difference to fish populations; and the measures needed to save Kaziranga's rhinos raise real concerns about the human costs of protection. Conservation must always worry about its shortcomings, be self-critical, and strive to do things better. But improving things as you go along is very different from not starting them at all (or giving up) because they're not perfect. As the need for conservation grows, we have to expand and further diversify what we do. And that will require having the courage to try new things, even though they're not (and may never be) perfect.

Great champions. Time and money. Big-picture thinking. Data that's fit for purpose. The imagination to broaden the argument. Political nous. And a belief that getting on with it now is better than waiting for perfection. These then, are the seven winning attributes I think are worth highlighting from the projects I've witnessed: ingredients for success that others working in conservation might be able to adapt to what they're doing. But conservation is too important to be left just to paid professionals, so just before leaving the question of how to make conservation more effective, it's worth thinking briefly about what ordinary people can do to help. The story of one man in particular offers inspiration.

People Power

Zhang Chunshan led a quiet life farming potatoes and corn in Yunnan Province, southwest China—until, that is, he came face to face with the sudden devastation by big business of the yew forests around his remote mountain home. The trees—whose bark was being removed in vast quantities to supply a recently approved anticancer drug called paclitaxel[1]—were supposedly protected by Chinese law. Yet neither the police nor forestry officials intervened to stop the frenzied harvest. Zhang alone decided to act. He went into the forests and confronted the bark poachers directly. He explained how the illegal gold rush was destroying the forest and, remarkably, persuaded many bark collectors to stop. His strategy began to work.

But for Yunnan Hande Bio Tech, the joint American-Chinese pharmaceutical company that hired the harvesters, refined the drug, and exported it to supply lucrative markets overseas, Zhang was becoming a problem. They responded with force. Zhang's house was attacked with stones, and he received anonymous death threats. His family grew increasingly fearful. Yet rather than give in, Zhang persisted. He investigated the company's

1. Paclitaxel is one of the most effective and widely used drugs for treating ovarian, lung, and breast cancers, but its uptake has come at a heavy price to the world's yew trees. On average it takes the bark from six 100-year-old trees to treat just one patient. Originally the drug was extracted from yews in the old-growth forests of the Pacific Northwest. By the mid-1990s environmental and commercial concerns led to the development of techniques for extracting a precursor to paclitaxel from the needles of the European yew. However, it remained cheaper to extract the drug directly from bark, so increased protection for Pacific yews led unscrupulous manufacturers to source the drug instead from Asian forests. Parts of Yunnan lost 90 percent of their yew trees, and across Asia destructive harvesting remains a major problem.

activities in more detail and handed over his findings to the provincial government. He went to the Chinese media, telling his story and that of the forests that he cared about to the newspapers and TV channels. The media publicized the problem, and Zhang persevered.

Eventually, 5 years after Zhang began the fight for his trees, a court in Yunnan's capital Kunming found the company guilty of illegally purchasing endangered plants, illegally processing them, and smuggling. They were fined US$1.2 million, the company's president was sentenced to 18 years in prison, and the yew-stripping operation was brought to an end. One individual had taken on a powerful and ruthless business and the inertia of a vast bureaucracy, and won.

And ordinary citizens, of course, can make conservation happen in less fraught circumstances too. Remember the proposal to build a marina in the heart of a much-loved wetland on the edge of my home town? Again, people didn't wait for officials to act. Ely's residents formed a campaign group; made allies with sailors, dog walkers, runners, and other worried users of the area; and took advice from professional conservation organizations. In the wake of people's outrage as trees were felled and trenches dug, we recruited 1,200 paid-up members.

Some of us collated data on the site's importance for conservation. Some learned more than is good for anyone about the minutiae of planning legislation. Artists auctioned their work, and cooks held cake sales to pay for lawyers.[2] Many wrote impassioned letters. And many felt a strengthened sense of belonging to their community.

Our information helped the government's national conservation advisors to include the area in a new Site of Special Scientific Interest,[3] and our arguments eventually prodded reluctant local bureaucrats to prosecute its owner for unauthorized development. Five years on and there is much that remains to be done to secure the long-term future of the whole of what we've started calling the Ely Wildspace. But its quiet waters are still free from flotillas of motorboats, and in their absence, kingfishers zip, electric-blue, along the bank, marsh harriers wheel over reedbeds where

2. At normal rates I reckon you need to sell around 50 sponge cakes to pay for an hour of a barrister's time. That's about 160,000 kilocalories' worth of baking per hour of advice. So it's just as well that at the end of it all the good folk of Ely had somewhere beautiful to walk.

3. Designating places as SSSIs gives them the greatest possible protection under UK law. In the case of the Ely Pits and Meadows SSSI, the listing means landowners are strongly discouraged from causing (and must put right) any significant harm they do to bitterns or breeding wetland birds—the species for which the site was notified. Designation also helps sympathetic owners access conservation grants; as a result several are now working with the local community to further improve the area for wildlife and for people.

near-invisible bitterns stand poised to spear fish, and more than ever, local people treasure the oasis on their doorstep.

And while Ely's battle of bittern impressed on me how much and how many people care about nature, you don't have to get involved in a campaign to have an impact. There are many other, equally powerful ways to make a difference, from letting elected officials know your views on the big issues to making informed choices about the products you buy. I've picked out 10 ideas and put them in an appendix at the back of the book. I hope some are useful and enable you to do more for nature.

But even if we all do more, even if conservation programs the world over scale up and increase their success rate, will that be enough? Is there a chance that most of the extraordinary fabric of the living world could then persist? Or is nature's continued demise utterly unavoidable?

What Hope Nature?

The short answer is that a great deal of nature could be saved. But will it be? That all depends on how much we want it, and on the decisions we make today and over the next quarter-century or so. While the projects I've visited show clearly that conservation can work—that things can get better—my list is also the result of deliberate cherry-picking. The sad reality is that the majority of conservation efforts are not so successful, and most activities that diminish nature—catching too many fish, cutting down a piece of forest or draining a wetland—evoke no organized response at all. But that said, in the time since Norman Moore took me to Stoborough Heath and I started drawing up my compendium of hope, I've realized there are very many more success stories than those I've focused on, and just like the ones I looked at they're not always where you might expect.

In China, for instance—portrayed by many a headline writer as an environmental bogeyman—some things are getting better. After decades of decline as a result of habitat loss and hunting, the number of wild giant pandas is starting to climb, and Chinese scientists are now looking at ways of restoring forests so as to reconnect isolated panda populations. The crested ibis—a sickle-billed, white-and-salmon-plumaged denizen of east Asian wetlands that was once one of the rarest birds on the planet—is now also beginning to recover. In 1981 there were only 7 known birds left in the wild; now there are over 500.

More broadly, China has gone further than just about any other country in addressing an underlying cause of biodiversity loss by limiting the size of its human population. Since the 1970s its controversial one-child policy has restricted almost all couples to having just a single son or daughter. The human costs are substantial and widely commented on in the West: the end of people's freedom to choose their family size, elevated rates of abortion and female infanticide, and worries about how relatively few working people will manage to support a rapidly ageing population—not to mention the sobering reality that an entire generation has no need for the words "brother" or "sister." But within China the policy has popular support, and its positive environmental effect is hard to question: whatever the eventual impact of 1.3 billion Chinese people becoming a good deal wealthier, imagine what it would be if there were hundreds of millions more. It strikes me that in some people's hypocritical rush to deplore the downsides of China's heady economic growth they conveniently overlook the fact that the Chinese have already made an immense sacrifice without which the world's environmental problems would be considerably greater.

Other developments are perhaps easier to cheer about. In the waters around the Antarctic, for example, 31 nations not always noted for working harmoniously have agreed and largely adhered to an international convention restricting their fishing operations.[4] The groundbreaking regulations place unprecedented emphasis on taking uncertainty into account when setting quotas and on ensuring we don't behave as if we were the only predator in the Southern Ocean but instead leave behind enough prey to sustain healthy populations of penguins, seals and albatrosses. Since the convention came into force Antarctica's fish and krill stocks have been maintained; a much-feared free-for-all has been averted—at least for now.

And in the Brazilian Amazon—a place where we're used to hearing about Belgium-sized areas of forest being cleared every year—deforestation rates plummeted between 2005 and 2010 to less than one-third of their long-term average.[5] It remains to be seen whether this reduction will be sustained as the global economy comes out of recession; much deforestation is driven by overseas demand for Amazon-reared beef and, to a lesser degree,

---

4. The agreement goes by the snappy title of the Convention on the Conservation of Antarctic Marine Living Resources, and its signatories include Argentina, Australia, France, Germany, Japan, Russia, the UK and the United States. Its main focus has been to avoid the collapse through overfishing of krill (not a fish at all, of course)—still one of the world's greatest living resources.

5. Clearance rates peaked in mid-1990s and early 2000s at around 0.9 Belgiums per year. By 2010 they were down to a 30-year low of around 0.2 Belgiums (or 2.5 Luxembourgs) per year. That's about 6,500 square kilometers per year in old money.

soybeans,[6] and rates have recently started to rise once again. But equally importantly, over the past decade forest protection has also increased spectacularly, so that around a half of all the Brazilian Amazon[7] is now designated as some form of reserve. Protected areas in this part of the world have been shown to slow habitat loss, so the hope is that even as markets recover, rates of forest loss might remain low—though that may to a large degree depend on how far the rest of the world is willing to pay for the carbon storage benefits it gets from Brazil foregoing further forest clearance.

Key to the Amazon story so far has been the pivotal role that both overseas and Brazilian companies involved in buying or bankrolling cattle and soy production have played in refusing to do business with operations that are still clearing forest. And new partnerships are behind another unlikely seeming conservation breakthrough, this time around the fringes of the Grand Canyon. Here the high-profile reintroduction of spectacular (and spectacularly rare) California condors has been jeopardized by the birds getting lead poisoning from deer entrails left behind by hunters. Once they hit flesh, lead bullets shatter into hundreds of fragments, lodging in guts and muscle. The only practical way to prevent accidental condor poisoning is for hunters to switch to copper bullets, but no less influential an organization than the National Rifle Association opposed such a change, arguing that lead makes for better shooting. The turning point came when wildlife biologists realized that if lead smithereens were poisoning condors they were probably finding their way into the venison on hunters' kids' dinner plates too. Now hunters are shown x-ray images of deer carcasses speckled with lead fragments and are voluntarily switching to copper ammunition—the power, once again, of thinking not like conservationists but like the people conservationists need to work with.

In truth, however, while conservation is winning some battles, even the doughtiest Pollyanna would have to agree that we are losing the war. The long-term enemies—habitat loss, overkill, and invasive aliens—are still with us, and in many places are gaining in strength and reach. And new threats are on the rise: the elimination of wildlife populations and the disruption of ecological communities by climate change, the massive buildup

---

6. Ironically, since the mid-1990s much Amazonian deforestation has been driven, in the wake of the bovine spongiform encephalopathy crisis, by demand from European and U.S. consumers for free-range beef, and by European Union restrictions on imports of genetically modified soy (unlike the U.S. crop, most Amazonian soy is GM free). Well-intended concerns have unintended consequences.

7. Or about 60 Belgiums . . .

of nitrogen compounds on land in the water, the spread of emerging infectious diseases, and the steady acidification of the world's oceans through absorption of the carbon dioxide we emit.

On top of all that, the ways we're threatening nature have the insidious habit of combining together so that the whole is greater than the sum of the parts. Some invaders, for instance, get better at out-competing the natives as habitats become degraded; the spread of others is enhanced by climate change. One poignant example of interacting threats comes from the islands of Hawaii, where habitat loss, nonnative predators, and accidentally introduced avian malaria have between them already restricted almost all the remaining natives to high altitude slopes. Now a warming climate is enabling invasive mosquitoes and the lethal malaria they carry to survive at progressively higher elevations: as they move up the mountains, they are likely to leave what remains of Hawaii's unique bird fauna with nowhere to go.

Many believe that the combined impact of all these threats is not simply to steadily diminish nature but to push it toward tipping points—thresholds beyond which ecosystems change fundamentally and irreversibly. In places such tipping points have already been exceeded. Persistent nutrient enrichment by runoff from fertilizer-drenched farmland has flipped previously fish-rich, nutrient-poor lakes into weed-rich, fish-free zones. Likewise, on the Grand Banks, the elimination of the dominant predator—adult cod— seems to have allowed the system to switch to one dominated instead by shrimps and snow crabs. Because the crabs eat baby cod, cod recovery looks unlikely, even though cod fishing has been banned there since 1992.

Scientists so far have only a sketchy understanding of how widespread such "quantum shifts" are, and where in different systems the thresholds lie. But even without such step changes, the consequences of nature's inexorable, grinding attrition are stark enough. Just ask the sad president of the dried-up *comuna* of Bambil, waiting for the water truck. A world where people have barely any wild fish left to catch. Where storms and floods go unabated by wetlands and where razing tropical forests that once soaked up our carbon has instead released billions of tonnes of greenhouse gases into the atmosphere. A world where Bambil-like stories of regret for what we've lost have become the norm.

Dismal for people, then, but for many other species the prognosis is even worse. Some creatures of course will do well. The wily and the weedy, the tolerant and the most adaptable: those that can make a living alongside people. But millions of other species are likely to become extinct. To be gone forever. Not hanging on, redeemable, in isolated patches of nature. Not even retained, with the eventual prospect of release, in captivity—there are far too few berths on the ark of the world's zoos and botanic gardens for that. But simply, irretrievably gone.

Like most conservationists I've asked, I think we have at most one generation left to avoid such an appalling prospect. In saying this I'm aware that similar statements have been made since the 1970s, yet while we've subsequently lost a great deal, the sky hasn't fallen in. But I'm also convinced that the situation now is far more serious than it was then. Since 1970 the human population has risen by over 80 percent. The global economy—as good a measure of the pressure on the natural world as any—has grown roughly fourfold. The finite nature of utterly vital natural resources for which we compete with other species (not niceties like oil or minerals, but essentials such as fresh water, land for growing food, and air that's clean enough to sustain life rather than cook or dissolve it) has become sharply apparent. Rates of loss of wild nature have accelerated; half of what once was has now been killed, cleared or converted. So with roughly 1 percent of what's left being eliminated each year, with the pressures we impose on the natural world still on the rise, and with the prospect of unforeseen interactions and tipping points, I'm in no doubt that if we carry on this way in a couple more decades the unraveling will have become irreparable.

Yet I also believe that we have it within our grasp to prevent such a catastrophe. Conservation business-as-usual will be nowhere near enough. Improving our success rate, in part perhaps through some of the ways I've picked out here—by having more resources, canny as well as wise leadership, better science, serious rather than rudimentary monitoring, and bolder, more innovative thinking—that will help. It will buy urgently needed time. But much more is necessary. To succeed, conservation has to scale up.

We already have a pretty good idea about the sorts of things that will be required on the ground and in the water: nature reserves that are bigger, better connected, and better protected; an end to those perverse government subsidies that encourage resource profligacy, and in their place intelligent combinations of sticks and carrots that drive much greater self restraint in how we exploit forests, fisheries, and the world's dwindling supplies of freshwater; far more serious efforts to limit the spread of invasive species; and a drastic reigning in of our nitrogen and greenhouse gas emissions.

Dramatic actions, but—lest we lapse into thinking them unaffordable—

things that are already happening in Costa Rica (with its carbon and water payments) and Loma Alta (where Don Alejandro and his fellow villagers voted to set aside 40 percent of their *comuna* for nature). Yet globally replicating and then sustaining these sorts of changes in the way we treat the environment is unlikely without much more fundamental transformations in how we live our lives: nothing quite as draconian as China's one-child policy, maybe, but bold and deeply challenging shifts nonetheless.

Having fewer children—especially in richer countries, where individuals consume so much more—is one of the basic changes we need to contemplate in the next 10 or 20 years if we're at all serious about sharing our planet with other species. So too is greatly reducing our use of fossil fuels to heat our homes and to travel for work and pleasure. And because producing animal protein generally uses far more land than growing grains or pulses, the wealthier among us need to seriously lower our consumption of meat and dairy if we're to meet rapidly rising food demand without disastrous consequences for other species. Having fewer cows farting methane—a far more powerful greenhouse gas than carbon dioxide—would help curb climate change too.

More generally, those of us—and there are many—who already have enough must seriously reconsider our seemingly insatiable appetite for yet more: the latest fashions, the cleverest gizmos, the shiniest appliances. Mortgaging the planet's future to meet basic human needs is one thing; doing it so we can consume yet more trivia is something else. Most fundamentally of all, many (myself included) believe we need to replace the prevailing global model of indefinite, resource-based economic growth. This may have been tenable in the nineteenth and early twentieth centuries, when one could be forgiven for thinking that the earth's resources were boundless. But to continue clinging to such a paradigm in today's small, hungry and so-evidently finite world seems dangerously naïve and desperately unimaginative.

These are major changes that challenge our core values and expectations and are unsettling even to think about, let alone adopt. If they happen at all, they'll take time. But I can think of three reasons for remaining optimistic—for believing that we still have world enough and time. Just.

The first is that these step changes are also essential for tackling the two other great crises of our age: human-driven alterations to the global climate, and the wretched poverty of many in the developing world. All three require those who have to live more modestly so that those who have not and those who are yet to come may have enough. If we are to succeed on climate change, on poverty or in conservation, then we need to start work now on addressing their common root causes.

Second, while nature is in very serious trouble, roughly half of it still remains: on average one-third of the shrublands (like fynbos and kwongan) found in regions with mediterranean climates, more than half of the great rainforests of the tropics, most of the coral reefs, and—though much less abundant than in the past—almost all of our fellow species. And this great, gritty persistence gives us time—not much, at the rate things are going, but a little bit.

Last, the places and people I've been lucky enough to encounter in researching this book have convinced me that we already have the wit and the will to slow down nature's loss very considerably: to buy more time to make the transition to sustainability. We know much of what we need to do. Often—more often than we might think—the will is there too. Whether it's Matiram Phukon's humbling tolerance of his elephant neighbors, the Alcoa miners wanting to be proud of their rehab work, or 1,200 of my neighbors signing up to save their bit of wetland, I've learned that many people really do want to do the right thing for nature. Together the stories I've looked at show that even over the last decade or two of unprecedented human pressure, when wit and will come together remarkable things can be achieved.

Dangerous but vulnerable animals with a high price on their heads have persisted despite being surrounded by grinding poverty and political instability. Corporations like Alcoa and countries like the Netherlands have invested immense amounts of money in conservation without it damaging their financial credibility. Strategic new alliances between fishermen, retailers, and conservationists have begun to reward sustainable fishing practices, even on the high seas. In the United States, smart changes to legislation have helped private landowners contribute to conservation. And in South Africa and Ecuador, restoring and protecting wild areas hasn't increased rural poverty—it's done the exact opposite.

Genuine progress, then, is being made. Around the world, conservation is succeeding. Its gains are of course fragmentary, and there is an immense amount more to be done. But by providing a basis for cautious optimism, places like Loma Alta and heroes like Norman Moore remind us that nature's unrelenting decline is not inevitable. Its conservation does not run counter to the wider economic interests of society at large, and instead enriches rather than diminishes the human enterprise. There can and will be more places like Stoborough, where habitat loss is stopped and then reversed. New and powerful forces for conservation—like those responsible businesses and investors that have slowed deforestation in the Amazon—will emerge. And brave individuals like Zhang Chunshan will speak out against the unscrupulous and unnecessary, and prevail. Everyone—layperson and professional, adult and child, consumer and voter—can help save a great deal of what is left. And although time is running out, there is still an enormous amount of nature left to fight for.

Nature's glass is still half full. There is hope for the wild.

APPENDIX: STEMMING THE LOSS

(Or What We Can All Do to Save Nature)

So what can we do? David MacKay, the UK government's top advisor on climate change and author of a wonderful book called *Sustainable Energy—Without the Hot Air*, is fond of pointing out that if everyone does a little, that's all, collectively, that we can hope to achieve: just a little. He's right of course. Keeping planetary carbon dioxide levels within safe margins isn't about changing the lightbulbs and unplugging the phone charger; it's about many, more fundamental and more challenging shifts in how we live. And the same goes for saving nature. We won't avert mass extinction by topping up the bird feeder. We need to achieve a great deal more, and quickly. Fortunately, there are many things that nearly all of us can do. I'm sure you have your own list of contributions you already make or things you feel you ought to do but haven't quite got round to yet. Here's mine.

1. Support the professionals. Perhaps the most obvious way to help is to give money to conservation organizations. And it's a fantastically useful thing to do: not— as some argue—just passing the buck. As I've found on my travels, conservation NGOs play many vital roles, from running projects on the ground that save species and safeguard habitats, to lobbying governments, raising public awareness, and chivvying big business into doing the right thing. But they can't do this without money—yours and mine.

There's a bewildering number of organizations to choose among, from the big international NGOs to national-level organizations and ones working on particular species or problems, through to local groups focusing in a specific area. My own strategy for donation is to find out as much as I can about what different NGOs do and what difference it makes (always being a bit skeptical about those claiming they achieve great things on their own), and then to support a handful, long term.

All the royalties from this book are going to three NGOs whose work and impact I've seen firsthand—Aaranyak (http://www.aaranyak.org), which works in Kaziranga and across northeast India; Life Net (http://www.lifenetnature.org), which runs small-scale projects at Loma Alta and elsewhere in the Americas and was founded by Dusti Becker; and Nature Conservation Foundation (http://www.ncf-india.org), which set up the Spiti Valley snow leopard project, among many others. Our donations can enable them—and hundreds of hardworking NGOs like them—to make a difference where it matters.

2. **Do it yourself.** Many people want to contribute to conservation more directly than just by giving money. Volunteers can help in all sorts of ways, from hands-on activities like habitat restoration and surveying animals and plants to fundraising and spreading the word about what's being done and why. Enthusiasm for physical labor (planting trees, cutting scrub, clearing ditches . . .) can be valuable, but often so are particular talents, like teaching, marketing, working with the media, and being able to maintain websites or databases. Many NGOs and projects greatly welcome such contributions and run well-organized volunteer programs. Sometimes, however, when no one else is tackling the problem you're passionate about, the only answer may be to start you own campaign. Usually the circumstances aren't as difficult as those Zhang Chunshan tackled in Yunnan. As I found out in Ely, a committed group with good local knowledge can get a lot done—especially when they think strategically and work with others who share their objectives.

3. **Let the powerful know.** Conservation is about taking the longer and the wider view—making decisions that benefit future as well as present generations, and that are in the interests of people—and life—as a whole. Sadly, this means it's not necessarily an approach that comes naturally to politicians, who need to deliver results in just three or four years for the people that put them in power. So if nature matters to you, it's vital you let your politicians know. Write to them about the issues you care about, ask for their support, and when it comes to elections find out where their parties stand—not just on the environment in the narrow sense, but in all areas of policy that affect it: on perverse subsidies, trade, transport, energy, education, and above all economic growth. And in the same way if you're a shareholder in a company, find out about their stance on the environment and if you don't like it, tell them. We can't afford to wait for politicians and businesspeople to lead us to a more sustainable world. Instead, those with votes or investments have a responsibility to push them in the right direction.

4. Buy carefully. As consumers we all have tremendous power to change people's impact on the environment, through the direct effects of what we (or our workplaces) buy, and via the signals that our purchasing patterns send to manufacturers and fellow consumers. Rather than focusing solely on conventional price tags we can look for items that cost less environmentally—in terms of how and where they're produced, how efficiently they operate, and how long they last. It's not that hard.

Fuel-efficient cars, low-energy appliances, garden plants that don't need extra watering, and so on: many products are now graded for their operating efficiency. Certification bodies like the Marine Stewardship Council (http://www.msc.org) and the Forest Stewardship Council (http://www.fsc.org) have become so well developed that in many countries there's really no decent excuse for buying unsustainably harvested fish or wood. If you want a nature-based holiday it's getting easier to pick operators that use community-owned guesthouses and guiding businesses —so more of your money goes to the local people whose attitudes will often determine the future of the wildlife you want to see. And there's a multitude of ways you can use your purchasing power to directly encourage sympathetic land management. South Africans can help honey badgers by buying honey from farmers who protect their hives better and so don't feel compelled to persecute the would-be hive-raiders; Chinese shoppers can buy rice grown without agrochemicals in the paddyfields that support the endangered crested ibis; health-conscious rooibos tea drinkers can now be nature conscious too and source their brew from a cooperative with impeccable low-impact credentials;[1] and anyone with a credit card and a penchant for ethnic knitwear can reward Mongolian herders for learning to live with snow leopards by buying their woolen slippers and rugs via the Snow Leopard Trust (http://www.snowleopard.org).

But it's not completely straightforward. For instance, the energy used to manufacture a car (its "embodied energy") is typically around one-sixth of all the energy used to drive it—so replacing your not-so-old vehicle with one that's only marginally more fuel-efficient probably does more harm than good. Likewise, buying pesticide-free organic food may be less environmentally beneficial than you think when it's grown far way (so transport costs are high) or at low yields (so that more land is needed for farming,

---

1. The post-apartheid boom in overseas demand for antioxidant-rich rooibos tea (made from the fynbos-endemic plant *Aspalathus linearis*) has triggered an explosion in the area under rooibos cultivation, which already threatens no fewer than 149 plant species with extinction. You can get guilt-free rooibos from the Heiveld Cooperative via http://www.equalexchange.coop/ and http://www.equalexchange.co.uk/.

leaving less for nature). And certification schemes vary in their standards and aims. Some concentrate on social benefits, others are environmentally focused, so which should you choose? Getting it exactly right can be tricky, but don't let that be an excuse for inaction. As a rule of thumb buying certified or energy-efficient items is almost certainly better than buying the alternatives, while for bigger purchases getting as much up-to-date information will probably help make a good purchase better still.

5. **Buy less.** Even more than buying better, we need to wean ourselves off buying so much. In arguing the case for a shift to greater sustainability the prescient American conservationist Aldo Leopold wrote, "Nothing could be more salutary at this stage than a little healthy contempt for a plethora of material blessings."[2] "This stage" was back in 1949, and experience since has served only to underscore the truth of Leopold's message.

Given that per capita income is rising so much more rapidly than the number of people doing the consuming, we all need to do what we can to cut back our demand for ever more stuff—state-of-the-art gadgets, fashionable furnishings, two-for-the-price-of-one accessories that hitherto we haven't needed at all. But consuming less is not just about consumer goods.

Kilo for kilo it takes between 3 and 15 times as much land (and generates 2 to 6 times more greenhouse gas) if our food comes from animals rather than plants,[3] so one of the easiest and most effective ways of reducing your environmental footprint is to eat less farm-reared meat and dairy produce. Have a meat-free day once a week (or better still, once a day). In a world of growing water scarcity where freshwater habitats and species are more threatened than any others, wasting less water is vital too—and simple steps (like collecting rainwater for the garden, swapping baths for showers, and flushing the toilet less often) can all help. And of course whether we're concerned about climate change or ocean acidification (or both), it's absolutely essential that we seriously lower our use of fossil fuels—by insulating our homes better, by being more accepting of winter cold and summer

2. From the foreword to *A Sand County Almanac* (1949). I sometimes wonder what the world might be like nowadays—socially and politically, as well as ecologically—if back then we'd paid more attention to Leopold's wisdom.

3. The variation is because it depends on what animal (or plant) you're talking about (poultry are much more efficient energy converters than pigs, which in turn outperform cows); whether you use weight, protein, or energy as the basis for your comparison; and the fact that some animals can be reared on waste products or on grassland that's unsuited to crop production. But regardless of the detail it's safe to say that eating less grain-fed meat would reduce pressure on dwindling natural resources.

heat, and by traveling less, more slowly, and by bus or train rather than car or plane.

Finally, if you've gone as far as you can toward becoming a low-energy, water-thrifty vegetarian at home, see what you can do in your workplace: many employers are becoming increasingly keen on cutting their footprint to save costs, keep up with the competition, and stay ahead of legislation.

6. Ease your guilt. Most of us are only human and so probably tread less lightly on the earth than we'd ideally like. My big indulgence—like many readers of this book, I suspect—is going to where the wild things are, preferably in the company of my family. And so even though we've become mostly vegetarian, our car doesn't move for days on end, and we're so thermostat-conscious that wintertime guests opt to keep their coats on, we still take a long-haul flight every couple of years to see the weird and the wonderful. Our compromise solution—not as good as quitting the habit but perhaps the next best thing—is to voluntarily pay a carbon tax to offset the emissions from our flights.

There are plenty of offset organizations to choose among (and most are happy to offset far more than just flights), but the key things to look for in picking one are what's called permanence (i.e., being reasonably certain that the carbon savings the scheme supports are likely to last); the amount of carbon dioxide the organization reckons your activities have emitted (easy to underestimate, so to be safe go for a more demanding scheme[4]); and whether the project being funded simultaneously delivers other benefits you care about (such as conservation or poverty alleviation). I've offset the travel involved in researching this book through a fund called Elemental Equity (http://www.elementalequity.org). They're investing in a remarkable South African project (devised by Richard Cowling and his colleagues, and explained a bit more in chapter 4) to restore carbon-thirsty thicket vegetation in the Baviaanskloof Nature Reserve. This not only soaks up carbon dioxide but recreates wildlife habitat, restores soil and provides badly needed jobs to former Working for Water employees into the bargain.

The cost of easing your carbon guilt this way is not negligible, but it's modest enough—and after all, if we're not willing to pay a hundred or so dollars extra on a trip costing thousands, perhaps we shouldn't be traveling in the first place.

---

4. There's an excellent emissions calculator—which adjusts for how a plane's fuel efficiency varies with flight distance, and takes into account its emissions of other greenhouse gases such as nitrous oxide—at http://www.atmosfair.de/.

Figure 26. Experimental thicket regeneration in the Baviaanskloof Nature Reserve, South Africa. The trials have demonstrated this is a cost-effective way to soak up carbon dioxide while simultaneously restoring species-rich habitat and providing jobs. Photo by Andrew Balmford.

7. Engage with change. Unfortunately, being serious about sustainability will increasingly require more than just cutting back when it's comfortable to do so and mitigating the damage when it isn't. As I argue in my final chapter, I think more drastic changes are inescapable. We must think seriously—especially in developed countries—about lowering our population by having smaller families. Given that many of our impacts depend more closely on the number of households than on total population size, getting on with the family we do have (separating from partners less often, and having adult offspring, parents and grandparents under one roof) is important too. We also have to confront some uncomfortable realities which will probably require difficult choices—like expanding nuclear power if we really want to avoid dangerous climate change, axing or redirecting subsidies that support ecologically unsustainable fishing or forestry (even when that costs jobs), and accepting further intensification of farming if we want to feed people properly without destroying much remaining natural habitat. And crucially, I believe we must start to switch to a different global model of economic performance—one that relies not on the self-evidently bankrupt model of indefinite economic expansion based on finite natural resources, but which instead has a more imaginative view of prosperity, based on social and natural capital rather than money and stuff.[5] Many of these shifts are radical challenges to our worldview—but to me they seem unavoidable if we genuinely want to avert nature's collapse (or address global poverty or stem dangerous climate change), and so the sooner we start discussing them seriously and with open minds the better.

8. Enjoy nature. Whatever the future holds, one of the most valuable things we can do for nature is simply spend time in it. Natural history can be a great entry point, but you certainly don't have to be a botanist or a hardcore birdwatcher to notice and to wonder at the living world. Whether it's in the Serengeti or a city park, nature has the capacity to inspire, unwind, intrigue, and humble. There's clear evidence that it's good for us too; studies from around the world show that having access to green space keeps adults fitter, increases children's concentration, coordination and self-discipline, lowers stress, reduces the incidence of crime and violent behavior, speeds up recovery from illness, and even helps us live longer. And getting out and enjoying nature aids in its conservation as well. By using areas where wildness hangs on we make landowners and politicians more aware that these places are valued, and by visiting them regularly we're much more likely to spot changes and threats which might require conservation action.

9. Help others go wild. Despite the manifest benefits of spending time in nature, urbanization, our switch toward working inside and our increasingly sedentary lifestyles all mean we're doing so less and less. There's a powerful feedback effect here too, with parents whose childhoods were largely indoors being less likely to spend time in the great outdoors with their own children. Growing numbers of health professionals, teachers, and psychologists consider that after 10,000 generations of people living in intimate, everyday contact with nature, this rapid "extinction of experience" in just a lifetime or two is a major contributor to rising levels of obesity, stress, and depression. And of course if fewer of us experience nature and view it as integral to our daily lives, it seems inevitable that fewer of us will stand up for its persistence: we can hardly be passionate about things we scarcely know. So if you already spend time wandering the woods, hiking in the hills or pottering along the shoreline, next time you go take a friend. Go with your family, take your neighbors. Take someone elderly or

---

5. For a great primer on what a steady-state economy might look like, see Jackson, *Prosperity Without Growth*.

ill, who might not otherwise be able to get out and feel the benefit. And go there with the most important people of all—take the children.

10. **Don't give up.** I think the last and most important thing we can all do for nature—and really the message of this entire book—is to not give up on it. Despite the toll we have already exerted, around half of wild nature still endures. Although the changes we need to bring about to secure its future are daunting, addressing them will simultaneously help us reduce poverty and limit climate change—the two other great challenges of our age. And even though collectively we sometimes act as if other people, generations and species don't matter that much, the stories here show that we're capable of doing a great deal better.

The deaf and blind American author and activist Helen Keller wrote, "Optimism is the faith that leads to achievement; nothing can be done without hope."[6] Nature requires us to have optimism. All of the conservation champions I've met have faced adversities and setbacks, and could have given up hope, could have lost confidence. The difference is that they didn't. And neither should we.

6. From *Optimism: An Essay*, p. 67.

ACKNOWLEDGMENTS

Writing a book is one of the hardest things I've ever done, and it would have been completely impossible without the generous help of many people. They variously informed the underlying ideas, made the exercise possible practically and financially, and provided the stories on which the book is based. I have been repeatedly humbled by their kindness, patience, and generosity and am deeply grateful to them all.

Above all, this project has all been inspired by three extraordinary conservationists and colleagues whom I feel privileged to count as friends: Richard Cowling, Rhys Green, and Norman Moore. Their enduring achievements bear tangible witness to the near-irresistible power of reason, passion, and hard work. My wonderful family—Sarah, Ben, and Jonah—have provided unstinting encouragement, a deep grounding in what really matters, and unending tolerance of all the time I've spent away or distracted from family life; it's my turn to walk the dog for the next six years. My parents nurtured my childhood passion for creatures, while Sue and Bob have been the most generous in-laws imaginable.

I am very grateful indeed to the Leverhulme Trust for essential support of a very different kind: a Research Fellowship, without which I would literally have gone nowhere. Friends and colleagues at the Department of Zoology in Cambridge have been astonishingly indulgent in allowing me to travel far and wide—and indeed to not travel at all but stay at home writing. I feel very lucky indeed to work where I do, and the support of Michael Akam, Malcolm Burrows, Julian Jacobs, and everyone in the Conservation Science Group is particularly appreciated. I'm very grateful indeed to the Chicago team, especially to my editor, Christie Henry, for her faith in the project and her immense patience during its long gestation, to Carrie Adams and Mary Gehl for applying their marketing and editing skills graciously and with good humor, and to Abby Collier for all her help in getting it over the finishing line. Judith Menes compiled the index efficiently and with professionalism. For sage advice at different stages of the project I wish to thank Tim Clutton-Brock, Jonathan Cobb, Rhys Green, Jim Kelly, Robert Macfarlane, Niall Mansfield, Mike McCarthy, Taylor Ricketts, Andrew Sugden, Jean Thomson Black, Rosie Trevelyan, and most especially my very kind and talented friend, Midge Gillies.

Each of the stories in the book has only been possible to write because of the time and kindness of a great many individuals. For talking with me or providing information about Kaziranga, I'm very grateful to Firoz Ahmed; Maan and Manju Barua; Polasz Bora; Pulin Bora; D. D. Boro; Namita Brahma; Suren Buragohain; Anawaruddin Choudhury; B. C. Chowdhury; Nigel Dudley; Martina di Fonzo; Biju Gogoi, Biju Hazarika; Nirupam Hazarika; Richard Kock; Shri Paramananda Lahan; Bibhuti Lahkar; M. C. Malakar; Robin Nath; Manish and Mina Patel; Bharat, Golap, and Holiram Patgiri; Matiram Phukon; Kees Rookmaaker; Bhimlal Saikia; Uttam Saikia; Pankaj Sharma; Bibhab Talukdar; Bhupen Talukdar; Jintu Tamuly; N. K. Vasu; and Belinda Wright.

Information on Safe Harbor came from talking with Michael Bean, Robert Bonnie, Jackie Britcher, Pete Campbell, Jay Carter III, Joe Chapman, Lamar Comalander, Ralph Costa, Jim Gray, Eric Holst, Connie Jess, Julian Johnson III, Brad Kocher, Norris Laffitte, Aaron Lange, Barclay McFaddin, Susan Miller, Julie Moore, Josh Raglin, Chris Storm, Pete van Hoorn, Dave Wilcove, and Bryan Zvolanek.

For all their input to the chapter on Working for Water, I'm grateful to Roland Black, William Bond, Paul Britton, Christopher Cowling, Richard Cowling, Caroline Gelderblom, Jennifer Gouza, Pierre Joubert, David le Maitre, Christo Marais, Edwill Moore, Ross Naude, Shirley Pierce, Mike Powell, Guy Preston, Tilla Raimondo, David Scott, Maggie and Peter Slingsby, Gerrid Umtwa, Katrina Vereen, and Brian van Wilgen.

My attempts to understand the Dutch National Ecological Network were greatly helped by Frank Alberts, Malcolm Ausden, Fred Baerselman, Hans Breeveld, Leen de Jong, Dirk Fey, Bart Fokkens, Jan Griekspoor, Monique Gulickx, Hans Kampf, Dick Landsmeer, Jacques van der Neut, Rob Rossel, Bill Sutherland, Janine van den Bos, Rob van der Werff, Mennobart van Eerden, Edwin van Oevelen, Bart van Tooren, Frans Vera, Vincent Wigbels, and Aart Zeeman.

In Ecuador, Evelyng Astudillo, Dusti Becker, Giovanny Catuto, Gregg Gorton, Tanner Johnson, Romelo Mendez, Leonardo Pozo, Don Alejandro Ramírez, Maria Luisa de Rumbea, Eugenio Tomalá, Virgilio Tomalá, Wilson Tomalá, Mauricio Torres, Pascual Torres, and the children of El Suspiro told me the remarkable story of the Reserva Ecológica de Loma Alta.

Bob Costanza, Amy Daniels, Paul Ferraro, Stefano Pagiola, Carlos Manuel Rodríguez and Arturo Sánchez-Azofeifa kindly let me quiz them about Costa's Rica's pioneering PSA program, while Ira and Peter Cooke introduced me to the wonderful world of WALFA.

For answering all my questions about Alcoa's restoration work in Western Australia I am very grateful to Glen Ainsworth, Dave Algar, Bradley "Buzzard" Ambrosius, John Asher, David Bell, Rob Brazell, Barry Carbon, Mike Craig, Paul de Tores, Kingsley Dixon, Royce Edwards, John Gardner, Al Glen, Carl Grant, Andrew Grigg, Larissa Hackett, Steve Hopper, Maureen Howard, John Koch, Norm McKenzie, Duncan Sutherland, Duarne Taaffe, Sharon Turner, Maree Weirheim, George White, and David Willyams.

I want to thank many AAFA members and friends for talking to me about their work and allowing me to sit in on their annual meeting: Andrew Bassford, Chip Bissell, Bobby Blocker, John Childers, Jody and Scott Hawkins, Craig Heberer, Carl Nish, Steve Rittenberg, Gary Sakagawa, Mike Shedora, Tim Thomas, Jack Vantress, Jack and Natalie Webster, and David Wilson. Insights into the workings of the Marine Stewardship Council came from conversations with Matt Bartholomew, Steve Broad, Scott Burns, Rupert Howes, Jim Humphreys, Jeremy Jackson, Jennifer Jacquet, Toby Middleton, Yemi Oloruntuyi, Daniel Pauly, André Punt, John Reynolds, Callum Roberts, Jeremy Ryland-Langley, Carl Safina, Amanda Stern-Pirlot, Mike Sutton, Jessica Wenban-Smith, and Kate Wilcox.

Many people kindly provided information on the various side stories in the book: Rob Ewers, Toby Gardner, and Carlos Souza helped me understand recent events in the Amazon; Andy Clarke and Inigo Everson took time to talk with me about how CCAMLR is improving the fate of fisheries around Antarctica; Li Bo, Zhang Chunshan, Andrew Laurie, and Yi Zhuangfang explained the fight for Yunnan's remaining yew trees; Kathy Hodder, Norman Moore, Nigel Symes, and Nigel Webb told me about the Dorset heathlands; Paul Butler sent me information on Rare's radio program in the western Pacific; M. D. Madhusudan and Charu Mishra talked to me about their work to save snow leopards in India's Spiti Valley; Richard Cowling, Mike Powell, and Marijn Zwinkels described how subtropical thicket restoration is soaking up carbon in South Africa; Chris Parish and Rhys Green explained their work on California condors; Chris Bowden, Binod Choudhury, Rhys Green, Debbie Pain, and Gerry Swan answered questions about the collapse of southern Asia's vulture population; and Andy Dobson and Gary Tabor shed light on the Yellowstone wolf project. Many other friends and colleagues helped by answering specific queries: Mark Avery, Justin Brashares, Mark Cocker, Jon Ekstrom, Marco Festa-

Bianchet, Brendan Fisher, Adrian Friday, Mac Hunter, Carl Jones, Val Kapos, Krithi Karanth, Nigel Leader-Williams, Aaron Lobo, Andrea Manica, Owen Mountford, Mark Mulligan, Thomasina Oldfield, Malvika Onial, Ben Phalan, Ana Rodrigues, Mark Spalding, Tim Sparks, Ed Tanner, Kerry Turner, Anton Weber, Sue White, and Lu Zhi.

Several people generously provided information on other success stories not covered here: Matthew Bell, Neil Burgess, Paul Butler, Ning Labbish Chao, Lauren Chapman, Curtis Freese, Rosemary Godfrey, Megan Hill, Terese Hor, Carl Jones, Tessa McGarry, Jean-Paul Paddack, Andy Plumptre, Dick Rice, Taylor Ricketts, Amanda Vincent, Ian Watson, Jeff Wielgus, and Tony Whitten. My failure to pursue their suggestions is entirely down to my limitations and constraints on the length of the book.

For doing a tremendous job with the maps I'm deeply grateful to Ruth Swetnam, and for providing the information on which they're based I thank Peter Cooke, Gilmar Navarrete Chacón, Cynthia Davis, Kathy Hodder, Susan Miller, Isabel Palacios, Carlos Manuel Rodríguez, Rob Rose, Stephen Sutton, Nigel Symes, Andrew Wannenburgh, and Penn State University (for providing the *Digital Chart of the World*). Peter Firth and Jonah Balmford very kindly edited the photos, which were generously provided by the American Albacore Fisheries Association, Paul Butler, Tom Collopy, Peter Cooke, Hans Kampf, Willy Metz, Charudutt Mishra, Norman Moore, Chris Parish, Shirley Pierce, Uttam Saikia, Kevin Smith, and Eric Spadgenske, with additional artwork by kind permission of Michael Edwards, Fabio Giovannoni, and Russ Malster. The references were put together with great professionalism by Ben Balmford, aided and abetted by Ben Phalan. Jon Green helped considerably with early background research.

Midge Gillies has spent many days helping to improve my writing, for which I'm profoundly grateful. As well as Midge, Sarah Blakeman, Andy Dobson, Rhys Green, Christie Henry, Taylor Ricketts, and Ruth Swetnam read every chapter and provided extremely helpful suggestions, while individual chapters have been read and commented on by Maan Barua, Dusti Becker, Robert Bonnie, Lamar Comalander, Peter Cooke, Richard Cowling, Mike Craig, Kingsley Dixon, William Foster, John Gardner, Monique Gulickx, Liz Hunter, Hans Kampf, Andrew Mallison, Susan Miller, Julie Moore, Norman Moore, Shirley Pierce, Carlos Manuel Rodríguez, Rob Rossel, Uttam Saikia, Pete van Hoorn, Brian van Wilgen, George White, and Kate Wilcox. This book is much better for all their help; any remaining errors are my fault entirely.

Andrew Balmford
Ely, July 2011

REFERENCES

Interviews

Except where stated, all interviews were conducted in person.

Chapter 1

Paul Butler, by telephone, September 2005; December 2005.
M. D. Madhusudan, Mysore, India, April 2010.
Charu Mishra, by e-mail, July 2010.
Norman Moore, Stoborough, UK, August 1999.
Nigel Symes, by telephone, September 2005.
Nigel Webb, by telephone, September 2005.

Chapter 2

Firoz Ahmed, Kaziranga, India, November 2007.
Maan Barua, Oxford, UK, October 2007.
Manju Barua, Kaziranga, India, November 2007.
Polash Bora, Kaziranga, India, November 2007.
Pulin Bora, Kaziranga, India, November 2007.
D. D. Boro, Kaziranga, India, November 2007.
Suren Buragohain, Kaziranga, India, November 2007.
B. C. Chowdhury, Dehradun, India, November 2007.
Martina di Fonzo, Cambridge, UK, October 2007.
Biju Gogoi, Kaziranga, India, November 2007.
Nirupam Hazarika, Guwahati, India, November 2007.
Shri Paramananda Lahan, Guwahati, India, November 2007.

Bibhuti Lahkar, Guwahati, India, November 2007.

M. C. Malakar, Guwahati, India, November 2007.

Robin Nath, Kaziranga, India, November 2007.

Golap Patgiri, Kaziranga, India, November 2007.

Holiram Patgiri, Kaziranga, India, November 2007.

Matiram Phukon, Kaziranga, India, November 2007.

Bhimlal Saikia, Kaziranga, India, November 2007.

Uttam Saikia, Kaziranga, India, November 2007.

Bibhab Talukdar, Guwahati, India, November 2007.

Bhupen Talukdar, Guwahati, India, November 2007.

N. K. Vasu, Dehradun, India, November 2007.

Belinda Wright, New Delhi, India, November 2007.

Chapter 3

Michael Bean, Washington DC, April 2008.

Robert Bonnie, Washington DC, April 2008.

Jackie Britcher, Southern Pines, North Carolina, April 2008.

Pete Campbell, Washington DC, April 2008.

Jay Carter III, Southern Pines, North Carolina, April 2008.

Joe Chapman, Brosnan Forest, North Carolina, April 2008.

Lamar Comalander, Brosnan Forest, North Carolina, April 2008.

Ralph Costa, by telephone, April 2008.

Jim Gray, Southern Pines, North Carolina, April 2008.

Eric Holst, Lodi, California, April 2008.

Connie Jess, Alameda County, California, April 2008.

Julian Johnson III, Southern Pines, North Carolina, April 2008.

Brad Kocher, Southern Pines, North Carolina, April 2008.

Norris Laffitte, by telephone, April 2008.

Aaron Lange, Mokelumne River, California, April 2008.

Barclay McFaddin, by telephone, April 2008.

Susan Miller, Southern Pines, North Carolina, April 2008.

Julie Moore, Washington DC, April 2008; by telephone, March 2009.

Josh Raglin, Brosnan Forest, North Carolina, April 2008.

Chris Storm, Mokelumne River, California, April 2008.

Pete van Hoorn, Alameda County, California, April 2008.

Dave Wilcove, by telephone, September 2006.

Bryan Zvolanek, Brosnan Forest, North Carolina, April 2008.

Chapter 4

Roland Black, Patensie, South Africa, April 2005.
Christopher Cowling, Cape St. Francis, South Africa, July 2005.
Richard Cowling, Cape St. Francis, South Africa, April 2005.
Caroline Gelderblom, Brasília, Brazil, July 2005.
Jennifer Gouza, Cape St. Francis, South Africa, April 2005.
Pierre Joubert, Patensie, South Africa, April 2005.
David le Maitre, Stellenbosch, South Africa, February 2005.
Edwill Moore, Patensie, South Africa, April 2005.
Ross Naude, Eersterivier, South Africa, May 2005.
Shirley Pierce, Cape St. Francis, South Africa, April 2005.
Mike Powell, Baviaanskloof, South Africa, April 2005.
Guy Preston, Cape Town, South Africa, February 2005.
Gerrid Umtwa, Langkloof, South Africa, April 2005.
Katrina Vereen, Patensie, South Africa, April 2005.
Brian van Wilgen, Stellenbosch, South Africa, February 2005; by telephone, June
    2005.

Chapter 5

Frank Alberts, Lelystad, The Netherlands, April 2007.
Malcolm Ausden, by telephone, October 2005.
Fred Baerselman, Oostvaarderplassen, The Netherlands, April 2007.
Hans Breeveld, Oostvaardersplassen, The Netherlands, April 2007.
Leen de Jong, Oostvaardersplassen, The Netherlands, April 2007.
Dirk Fey, Biesbosch, The Netherlands, April 2007.
Bart Fokkens, Lelystad, The Netherlands, April 2007.
Jan Griekspoor, Oostvaardersplassen, The Netherlands, April 2007.
Hans Kampf, Soest, The Netherlands, April 2007.
Jacques van der Neut, Biesbosch, The Netherlands, April 2007.
Rob Rossel, Hilversum, The Netherlands, April 2007.
Janine van den Bos, Lelystad, The Netherlands, April 2007.
Rob van der Werff, Lelystad, The Netherlands, April 2007.
Mennobart van Eerden, Lelystad, The Netherlands, April 2007.
Bart van Tooren, 's-Graveland, The Netherlands, April 2007.
Frans Vera, Oostvaardersplassen, the Netherlands, April 2007.

Vincent Wigbels, Lelystad, the Netherlands, April 2007.
Aart Zeeman, Hilversum, the Netherlands, April 2007.

Chapter 6

Evelyng Astudillo, Loma Alta, Ecuador, December 2007.
Dusti Becker, El Suspiro, Ecuador, December 2007.
Giovanny Catuto, Loma Alta, Ecuador, December 2007.
Ira Cooke, Cambridge, UK, May 2008.
Peter Cooke, by telephone, January 2010.
Bob Costanza, by e-mail, August 2007 .
Amy Daniels, by e-mail, August 2007.
Paul Ferraro, by e-mail, April 2006.
Romelo Mendez, Loma Alta, Ecuador, December 2007.
Stefano Pagiola, by e-mail, August 2007.
Leonardo Pozo, Loma Alta, Ecuador, December 2007.
Don Alejandro Ramírez, Loma Alta, Ecuador, December 2007.
Carlos Manuel Rodríguez, by telephone, December 2009.
Maria Luisa de Rumbea, Guayaquil, Ecuador, December 2007.
Arturo Sánchez-Azofeifa, by e-mail, December 2010.
Eugenio Tomalá, Bambil, Ecuador, December 2007.
Virgilio Tomalá, Loma Alta, Ecuador, December 2007.
Wilson Tomalá, Loma Alta, Ecuador, December 2007.
Mauricio Torres, El Suspiro, Ecuador, December 2007.
Pascual Torres, El Suspiro, Ecuador, December 2007.

Chapter 7

Glen Ainsworth, Huntly mine, Australia, December 2006.
Dave Algar, Dwellingup, Australia, December 2006.
Bradley "Buzzard" Ambrosius, Huntly mine, Australia, December 2006.
John Asher, Dwellingup, Australia, December 2006.
David Bell, by telephone, December 2006.
Barry Carbon, Perth, Australia, December 2006.
Mike Craig, Huntly mine, Australia, December 2006.
Paul de Tores, Dwellingup, Australia, December 2006.
Kingsley Dixon, Perth, Australia, December 2006.
Royce Edwards, Huntly mine, Australia, December 2006.

John Gardner, Huntly mine and Perth, Australia, December 2006.
Al Glen, Dwellingup, Australia, December 2006.
Carl Grant, Huntly mine, Australia, December 2006.
Steve Hopper, Cape St. Francis, South Africa, July 2005.
Maureen Howard, Huntly mine, Australia, December 2006.
John Koch, Huntly mine, Australia, December 2006.
Norm McKenzie, Perth, Australia, December 2006.
Duncan Sutherland, Dwellingup, Australia, December 2006.
Duarne Taaffe, Huntly mine, Australia, December 2006.
Sharon Turner, Huntly mine, Australia, December 2006.
George White, Boyanup, Australia, December 2006.

Chapter 8

Matt Bartholemew, London, February 2010.
Andrew Bassford, San Diego, California, April 2008.
Chip Bissell, San Diego, California, April 2008.
Bobby Blocker, San Diego, California, April 2008.
Steve Broad, London, September 2007.
Scott Burns, by telephone, November 2006; Annapolis, VA, April 2008.
John Childers, San Diego, California, April 2008.
Scott Hawkins, San Diego, California, April 2008.
Craig Heberer, San Diego, California, April 2008.
Rupert Howes, London, February 2010.
Jim Humphreys, San Diego, California, April 2008.
Jeremy Jackson, Cambridge, UK, May 2008.
Jennifer Jacquet, Cambridge, UK, February 2009.
Toby Middleton, London, February 2010.
Carl Nish, San Diego, California, April 2008.
Yemi Oloruntuyi, London, February 2010.
André Punt, by e-mail, February 2010.
John Reynolds, by e-mail, February 2010.
Steve Rittenberg, San Diego, California, April 2008.
Callum Roberts, by telephone, July 2009.
Jeremy Ryland-Langley, by telephone, February 2010.
Carl Safina, Cambridge, UK, March 2007.
Gary Sakagawa, San Diego, California, April 2008.
Mike Shedora, San Diego, California, April 2008.
Amanda Stern-Pirlot, London, February 2010.

Mike Sutton, Monterey, California, April 2008.

Tim Thomas, San Diego, California, April 2008.

Jack Vantress, San Diego, California, April 2008.

Jack Webster, San Diego, California, April 2008.

Natalie Webster, San Diego, California, April 2008.

Jessica Wenban-Smith, London, March 2008.

Kate Wilcox, London, February 2010.

David Wilson, San Diego, California, April 2008.

Chapter 9

Li Bo, by e-mail, December 2010.

Zhang Chunshan, by e-mail, December 2010.

Andy Clarke, Cambridge, UK, November 2005.

Inigo Everson, Cambridge, UK, March 2010.

Chris Parish, Vermillion Cliffs, Arizona, July 2011.

Printed Material

Below are the main sources I've used, organized by chapter and with an emphasis on peer-reviewed journals and books.

Chapter 1

Alexander, J. 1920. "Nobel Award to Haber." *New York Times*, February 3.

Balmford, A., A. Bruner, P. Cooper, et al. 2002. "Economic Reasons for Conserving Wild Nature." *Science* 297:950–53.

Barnosky, A. D., N. Matzke, S. Tomiya, et al. 2011. "Has the Earth's Sixth Mass Extinction already Arrived?" *Nature* 471:51–57.

Barua, M., and P. Jepson. 2010. "The Bull of the Bog: Bittern Conservation Practice in a Western Bio-Cultural Setting." In *Ethno-Ornithology: Birds, Indigenous Peoples, Culture and Society*, edited by Sonia Tidemann and Andrew Gosler, 301–12. London: Earthscan.

Baum, J. K., R. A. Myers, D. G. Kehler, B. Worm, S. J. Harley, and P. A. Doherty. 2003. "Collapse and Conservation of Shark Populations in the Northwest Atlantic." *Science* 299:389–92.

Berger, L., R. Speare, P. Daszak, et al. 1998. "Chytridiomycosis Causes Amphibian

Mortality Associated with Population Declines in the Rain Forests of Australia and Central America." *Proceedings of the National Academy of Sciences* 95:9031–36.

Butchart, S. H. M., A. J. Stattersfield, and N. J. Collar. 2006. "How Many Bird Extinctions Have We Prevented?" *Oryx* 40:266–78.

Butchart, S. H. M., M. Walpole, B. Collen, et al. 2010. "Global Biodiversity: Indicators of Recent Declines." *Science* 328:1164–68. doi:10.1126/science.1187512.

Chapron, G. 2005. "Re-Wilding: Other Projects Help Carnivores Stay Wild." *Nature* 437:318.

Charlson, R. J., J. E. Lovelock, M. O. Andreae, and S. G. Warren. 1987. "Oceanic Phytoplankton, Atmospheric Sulphur, Cloud Albedo and Climate." *Nature* 326:655–61.

Costanza, R., O. Pérez-Maqueo, M. L. Martinez, P. Sutton, S. J. Anderson, and K. Mulder. 2008. "The Value of Coastal Wetlands for Hurricane Protection." *Ambio* 37:241–48. doi:10.1579/0044-7447(2008)37[241:TVOCWF]2.0.CO;2.

Craigie, I. D., J. E. M. Baillie, A. Balmford, et al. 2010. "Large Mammal Population Declines in Africa's Protected Areas." *Biological Conservation* 143:2221–28.

Curtin, C. G. 2002. "Integration of Science and Community-Based Conservation in the Mexico/ U.S. Borderlands." *Conservation Biology* 16:880–86.

Defense Mapping Agency. 1992. *Digital Chart of the World*. Fairfax, VA: National Imagery and Mapping Agency.

Diaz, R. J, and R. Rosenberg. 2008. "Spreading Dead Zones and Consequences for Marine Ecosystems." *Science* 321:926–29.

Food and Agriculture Organization of the United Nations. 2010. *The State of World Fisheries and Aquaculture 2010*. Rome: FAO.

Fortin, D., H. L. Beyer, M. S. Boyce, D. W. Zhang, T. Duchesne, and J. S. Mao. 2005. "Wolves Influence Elk Movements: Behavior Shapes a Trophic Cascade in Yellowstone National Park." *Ecology* 86:1320–30.

Hamilton, A., A. Cunningham, D. Byarugaba, and F. Kayanja. 2000. "Conservation in a Region of Political Instability: Bwindi Impenetrable Forest, Uganda." *Conservation Biology* 14:1722–25.

Hardy, T. 1983. *The Return of the Native*. Harmondsworth, UK: Penguin. First published 1878 by Smith, Elder and Co.

Hoegh-Guldberg, O., P. J. Mumby, A. J. Hooten, et al. 2007. "Coral Reefs under Rapid Climate Change and Ocean Acidification." *Science* 318:1737–42.

Interacademy Panel on International Issues. 2009. *Interacademy Panel on International Issues Statement on Ocean Acidification*. Trieste, Italy: IAP.

International Union for Conservation of Nature. 2011. "The IUCN Red List of Threatened Species." Version 2011.1. http://www.iucnredlist.org/.

Jackson, J. B. C. 1997. "Reefs since Columbus." *Coral Reefs* 16:23–32.

Jackson, R. M., C. Mishra, T. M. McCarthy, and S. B. Ale. 2010. "Snow Leopards:

Conflict and Conservation." In *Biology and Conservation of Wild Felids*, edited by D. W. Macdonald and A. J. Loveridge, 419–34. Oxford: Oxford University Press.

Jones, C. G., W. Heck, R. E. Lewis, Y. Mungroo, G. Slade, and T. Cade. 1995. "The Restoration of the Mauritius Kestrel *Falco punctatus* Population." *Ibis* 137:S173–80.

Joppa, L. N., D. L. Roberts, and S. L. Pimm. 2011. "How Many Species of Flowering Plants Are There?" *Proceedings of the Royal Society B: Biological Sciences* 278:554.

Kauffman, M. J., J. F. Brodie, and E. S. Jules. 2010. "Are Wolves Saving Yellowstone's Aspen? A Landscape-Level Test of a Behaviorally Mediated Trophic Cascade." *Ecology* 91:2742–55.

Klein, A. M., B. E. Vaissiere, J. H. Cane, et al. 2007. "Importance of Pollinators in Changing Landscapes for World Crops." *Proceedings of the Royal Society B: Biological Sciences* 274:303–13.

Lee, R. 2011. "The Outlook for Population Growth." *Science* 333: 569–73.

Lotze, H. K., and B. Worm. 2009. "Historical Baselines for Large Marine Animals." *Trends in Ecology and Evolution* 24:254–62.

Manning, A. D., I. J. Gordon, and W. J. Ripple. 2009. "Restoring Landscapes of Fear with Wolves in the Scottish Highlands." *Biological Conservation* 142:2314–21.

Markandya, A., T. Taylor, A. Longo, et al. 2008. "Counting the Cost of Vulture Decline—an Appraisal of the Human Health and Other Benefits of Vultures in India." *Ecological Economics* 67:194–204.

McClenachan, L., J. B. C. Jackson, and M. J. H. Newman. 2006. "Conservation Implications of Historic Sea Turtle Nesting Beach Loss." *Frontiers in Ecology and the Environment* 4:290–96.

Millennium Ecosystem Assessment. 2005. *Ecosystems and Human Well-Being: Synthesis*. Washington DC: Island Press.

Miller, J. R. 2005. "Biodiversity Conservation and the Extinction of Experience." *Trends in Ecology & Evolution* 20:430–34.

Mishra, C., P. Allen, T. McCarthy, M. D. Madhusudan, A. Bayarjargal, and H. H. T. Prins. 2003. "The Role of Incentive Programs in Conserving the Snow Leopard." *Conservation Biology* 17:1512–20.

Moore, N. W. 1962. "The Heaths of Dorset and Their Conservation." *Journal of Ecology* 50:369–91.

Myers, R. A., J. K. Baum, T. D. Shepherd, S. P. Powers, and C. H. Peterson. 2007. "Cascading Effects of the Loss of Apex Predatory Sharks from a Coastal Ocean." *Science* 315:1846–50.

Oaks, J. L., M. Gilbert, M. Z. Virani, et al. 2004. "Diclofenac Residues as the Cause of Vulture Population Decline in Pakistan." *Nature* 427:630–33. doi:10.1038/nature02317.

Orr, J. C., V. J. Fabry, O. Aumont, et al. 2005. "Anthropogenic Ocean Acidification over the Twenty-First Century and Its Impact on Calcifying Organisms." *Nature* 437:681–86.

Pain, D. J., C. G. R. Bowden, A. A. Cunningham, et al. 2008. "The Race to Prevent the Extinction of South Asian Vultures." *Bird Conservation International* 18:S30–48.

Pain, D. J., A. A. Cunningham, P. F. Donald, et al. 2003. "Causes and Effects of Temporospatial Declines of *Gyps* Vultures in Asia." *Conservation Biology* 17:661–71.

Pyle, R. M. 2003. *The Thunder Tree: Lessons from a Secondhand Landscape*. Chicago: Houghton Mifflin.

Rodrigues, A. S. L. 2006. "Are Global Conservation Efforts Successful?" *Science* 313:1051–52.

Roosevelt, T. 1910. *The New Nationalism*. New York: Outlook Company.

Rose, R. J., N. R. Webb, R. T. Clarke, and C. H. Traynor. 2000. "Changes on the Heathlands in Dorset, England, between 1987 and 1996." *Biological Conservation* 93:117–25.

Schulte, P., L. Alegret, I. Arenillas, et al. 2010. "The Chicxulub Asteroid Impact and Mass Extinction at the Cretaceous-Paleogene Boundary." *Science* 327: 1214–18.

Sinervo, B., F. Méndez-de-la-Cruz, D. B. Miles, et al. 2010. "Erosion of Lizard Diversity by Climate Change and Altered Thermal Niches." *Science* 328:894–99.

Smith, D. W., R. O. Peterson, and D. B. Houston. 2003. "Yellowstone after Wolves." *BioScience* 53:330–40.

Sukhdev, P. 2010. "Biodiversity Protection Can Help Tackle Climate Change and Poverty." *Guardian*, June 10. http://www.guardian.co.uk/environment/cif-green/2010/jun/10/biodiversity-climate-change-poverty.

Sutton, M. A., O. Oenema, J. W. Erisman, A. Leip, H. Van Grinsven, and W. Winiwarter. 2011. "Too Much of a Good Thing." *Nature* 472:159–61.

Thomas, C. D., A. Cameron, R. E. Green, et al. 2004. "Extinction Risk from Climate Change." *Nature* 427:145–48.

Thomas, J. A. 1989. "Return of the Large Blue Butterfly." *British Wildlife* 1:2–13.

Towns, D. R., and K. G. Broome. 2003. "From Small Maria to Massive Campbell: Forty Years of Rat Eradications from New Zealand Islands." *New Zealand Journal of Zoology* 30:377–98.

Trivedi, B. P. 2001. "India Vulture Die-Off Spurs Carcass Crisis." *National Geographic News*, December 28. http://news.nationalgeographic.com/news/2001/12/1226_TVvulturedieoff.html.

United Nations Environment Programme. 2002. *UNEP Global Environment Outlook 3*. London: Earthscan.

Vaughan, P. W., A. Regis, and E. St. Catherine. 2000. "Effects of an Entertainment-Education Radio Soap Opera on Family Planning and HIV Prevention in St. Lucia." *International Family Planning Perspectives* 26:148–57.

Vucetich, J. A., D. W. Smith, and D. R. Stahler. 2005. "Influence of Harvest, Climate and Wolf Predation on Yellowstone Elk, 1961–2004." *Oikos* 111:259–70.

Weber, B., and A. Vedder. 2001. *In the Kingdom of Gorillas: Fragile Species in a Dangerous Land*. New York, NY: Simon & Schuster.

White, P. J., and R. A. Garrott. 2005. "Yellowstone's Ungulates after Wolves—Expectations, Realizations, and Predictions." *Biological Conservation* 125:141–52.

Wilmers, C. C., and W. M. Getz. 2005. "Gray Wolves as Climate Change Buffers in Yellowstone." *PLoS Biology* 3:0571–76.

WWF International, Institute of Zoology, and Global Footprint Network. 2010. *Living Planet Report 2010*. Gland, Switzerland; London; and Oakland, CA: WWF International, Institute of Zoology, and Global Footprint Network.

Chapter 2

Ahmad Zafir, A. W., J. Payne, A. Mohamed, et al. 2011. "Now or Never: What Will It Take to Save the Sumatran Rhinoceros *Dicerorhinus sumatrensis* from Extinction?" *Oryx* 45:225–33. doi:10.1017/S0030605310000864.

Barua, M., and P. Sharma. 1999. "Birds of Kaziranga National Park, India." *Forktail* 15:47–60.

Bhatt, S. 2005. *Opportunities and Limitations for Benefit Sharing in Select World Heritage Sites*. Delhi: UNESCO, IUCN, and WII.

Breeden, S., and B. Wright. 1996. *Through the Tiger's Eyes: A Chronicle of Indian Wildlife*. Berkeley, CA: Ten Speed Press.

Brener, A. E. 1998. "An Anti-Poaching Strategy for the Greater One-Horned Rhinoceros in Kaziranga National Park, Assam, India." Master's thesis, University of Calgary.

Check, E. 2006. "The Tiger's Retreat." *Nature* 441:927–30.

Choudhury, A. 2004. "Human-Elephant Conflicts in Northeast India." *Human Dimensions of Wildlife* 9:261–70.

Di Fonzo, M. M. I. 2007. "Determining Correlates of Human-Elephant Conflict Reports within Fringe Villages of Kaziranga National Park, Assam." Master's thesis, Imperial College London.

Gee, E. P. 1948a. "The Great Indian One-Horned Rhinoceros." *Zoo Life* 3:106–7.

———. 1948b. "The Great One-Horned Rhinoceros in Kaziranga Sanctuary, Assam." *Journal of the Bengal Natural History Society* 23:63–65.

Goswami, D. C., and P. J. Das. n.d. "Hydrological Impact of Earthquakes on the Brahmaputra river Regime, Assam: A Study in Exploring Some Evidences." *My Green Earth* 3, no. 2: 3–12.

Groves, C. P., P. Fernando, and J. Robovský. 2010. "The Sixth Rhino: A Taxonomic

Re-Assessment of the Critically Endangered Northern White Rhinoceros." *PLoS ONE* 5:e9703. doi:10.1371/journal.pone.0009703.

Mathur, V. B., A. Verma, N. Dudley, S. Stolton, M. Hockings, and R. James. ND. *Opportunities and Challenges for Kaziranga National Park, Assam.* UNF-UNESCO.

Menon, V. 1996. *Under Siege: Poaching and Protection of Greater One-Horned Rhinoceroses in India.* Cambridge: TRAFFIC International.

Misra, M. K. 2005. *Kaziranga National Park World Heritage Site: Review of Protection and Suggestions to Enhance Their Effectiveness: Development of a Comprehensive Capacity Building Plan for Frontline Staff.* Delhi: UNESCO, IUCN, and WII.

Monirul Qader Mirza, M., R. A. Warrick, N. J. Ericksen, and G. J. Kenny. 2001. "Are Floods Getting Worse in the Ganges, Brahmaputra and Meghna Basins?" *Global Environmental Change Part B: Environmental Hazards* 3:37–48.

Rehman, T. 2008. "Rhino Killings: The Inside Story." *Tehelka* 5, no. 5.

Rookmaaker, K. 2003. "Historical Records of the Sumatran Rhinoceros in North-East India." *Rhino Foundation for Nature in North East India Newsletter* 5:11–12.

Talukdar, B. K. 2000. "The Current State of Rhino in Assam and Threats in the 21st Century." *Pachyderm* 29 (July–December): 39–47.

———. 2002. "Dedication Leads to Reduced Rhino Poaching in Assam in Recent Years." *Pachyderm* 33 (July–December): 58–63.

———. 2003. "Importance of Anti Poaching Measures towards Successful Conservation and Protection of Rhinos and Elephants, North-Eastern India." *Pachyderm* 34 (January–June): 59–65.

———. 2006. "Assam Leads in Conserving the Greater One-Horned Rhinoceros in the New Millennium." *Pachyderm* 41 (July–December): 85–89.

Talukdar, B. K., M. Barua, and P. K. Sarma. 2007. "Tracing Straying Routes of Rhinoceros in Pabitora Wildlife Sanctuary, Assam." *Current Science* 92:1303–5.

Talukdar, B. K., and P. K. Sarma. 2007. *Indian Rhinos in Protected Areas of Assam: A Geo-Spatial Documentation of Habitat Changes and Threats.* Guwahati: Aaranyak.

Whitley, E. 1992. *Gerald Durrell's Army.* London: John Murray.

Chapter 3

Anonymous. 2008. "Doubly Endangered." *Nature* 454:1029.

Bailey, M. A. 2002. "A Gopher Tortoise Conservation Plan." *Alabama Wildlife*, Summer, 20–22.

Bean, M. J. 2004. "Rethinking Conservation Strategies after Thirty Years." In *Red-Cockaded Woodpecker: Road to Recovery*, edited by R. Costa and S. J. Daniels, 34–36. Surrey, BC: Hancock House.

―――. 2005. *The Endangered Species Act, Success or Failure?* Washington DC: Environmental Defense.

Bean, M. J., and L. E. Dwyer. 2000. "Mitigation Banking as an Endangered Species Conservation Tool. *Environmental Law Reporter News and Analysis* 30: 10537–56.

Bean, M. J., J. P. Jenny, and B. van Eerden. 2001. "Safe Harbor Agreements Carving Out a New Role for NGOs." *Conservation Biology in Practice* 2, no. 2: 8–16.

Benjamin, D. K. 2003. "Preemptive Cuts." *P.E.R.C. Reports* (December): 17–18.

Boersma, P. D., P. Kareiva, W. F. Fagan, J. A. Clark, and J. M. Hoekstra. 2001. "How Good Are Endangered Species Recovery Plans?" *BioScience* 51: 643. doi:10.1641/0006–3568(2001)051[0643:HGAESR]2.0.CO;2.

Bonnie, R. 1997. "Safe Harbor for the Red-Cockaded Woodpecker." *Journal of Forestry* 95, no. 4: 17–22.

―――. 1999. "Endangered Species Mitigation Banking: Promoting Recovery through Habitat Conservation Planning under the Endangered Species Act." *Science of the Total Environment* 240:11–19.

―――. 2002. "Safe Harbor for Species and Landowners." *Tree Farmer*, May-June, 6–9.

―――. 2004. "From Cone's Folly to Brosnan Forest and Beyond: Protecting Red-Cockaded Woodpeckers on Private Lands." In *Red-Cockaded Woodpecker: Road to Recovery*, edited by R. Costa and S. J. Daniels, 163–73. Surrey, BC: Hancock House.

―――. 2005. *Building on Success: Improving the Endangered Species Act*. Washington DC: Environmental Defense.

Britcher, J. J. 2006. "Woodpeckers Find a Home at Fort Bragg." *Endangered Species Bulletin* 31, no. 2: 28–30.

Chadwick, A. N. 2004. "South Carolina's Safe Harbor Program for Red-Cockaded Woodpeckers." In *Red-Cockaded Woodpecker: Road to Recovery*, edited by R. Costa and S. J. Daniels, 180–184. Surrey, BC: Hancock House.

Craftsman. 1996. "Developer's Guide to Endangered Species Regulation." Washington DC: Home Builder Press.

Drake, D., and E. J. Jones. 2002. "Forest Management Decisions of North Carolina Landowners Relative to the Red-Cockaded Woodpecker." *Wildlife Society Bulletin* 30: 121–30.

―――. 2003. "Current and Future Red-Cockaded Woodpecker Habitat Availability on Non-Industrial Private Forestland in North Carolina." *Wildlife Society Bulletin* 31:661–69.

Earley, L. S. 1996. "Safe Harbor in the Sandhills." *Wildlife in North Carolina*, October, 10–15.

―――. 2004. *Looking for Longleaf: The Fall and Rise of an American Forest*. Chapel Hill: University of North Carolina Press.

———. 2006. *A Working Forest*: rev. ed. Southern Pines, NC: Sandhills Area Land Trust.

Eilperin, J. 2005. "Landowners Split on Species Act Burden." *Washington Post*, October 18.

Environmental Defense. 2000. *Progress on the Back Forty: An Analysis of Three Incentive-Based Approaches to Endangered Species Conservation on Private Land.* Washington DC: Environmental Defense.

Fox, J., and A. Nino-Murcia. 2005. "Status of Species Conservation Banking in the United States." *Conservation Biology* 19:996–1007.

James, F. C. 1995. "Status of the Red-Cockaded Woodpecker and Its Habitat." In *Red-Cockaded Woodpecker, Recovery, Ecology and Management.*, edited by D. L. Kulhavy, R. G. Hooper, and R. Costa, 437–38. Texas: Center for Applied Studies in Forestry, Stephen F. Austin University.

Leopold, A. 1947. "The Ecological Consequences." In *The River of the Mother of God and Other Essays by Aldo Leopold*, edited by S. L. Flader and J. B. Callicot, 338–46. Madison: University of Wisconsin Press.

Lueck, D., and J. A. Michael. 2003. "Preemptive Habitat Destruction under the Endangered Species Act." *Journal of Law and Economics* 46: 27–60.

Luoma, J. R. 2001. "Safe Harbor." *Wildlife Conservation* 104, no. 3: 31–35, 65.

Male, T. D., and M. J. Bean. 2005. "Measuring Progress in US Endangered Species Conservation." *Ecology Letters* 8: 986–92.

Miller, S. L., P. V. Campbell, and M. A. Cantrell. 2004. "North Carolina Sandhills Red-Cockaded Woodpecker Safe Harbor Programme: Current Status and Lessons Learned." In *Red-Cockaded Woodpecker: Road to Recovery*, edited by R. Costa and S. J. Daniels, 174–79. Surrey, BC: Hancock House.

Schwab, I. R. 2002. "Cure for a Headache." *British Journal of Ophthalmology* 86:843.

Stone, R. 1995. "Incentives Offer Hope for Habitat." *Science* 269:1212–13.

Twain, M. 1867. *The Celebrated Jumping Frog of Calaveras County and Other Stories.* New York: C. H. Webb.

Walters, J. R., K. Brust, J. H. Carter III, and S. Anchor. 2009. *Red-Cockaded Woodpecker Demographic Monitoring in the North Carolina Sandhills: Is Safe Harbor a Biological Success?* Asheville, NC: Report to U.S. Fish and Wildlife Service.

Wilcove, D. S. K., M. J. Bean, R. Bonnie, and M. McMillan. 1996. *Rebuilding the Ark: Toward a More Effective Endangered Species Act for Private Land.* Washington DC: Environmental Defense Fund.

Wilcove, D. S., and J. Lee. 2004. "Using Economic and Regulatory Incentives to Restore Endangered Species: Lessons Learned from Three New Programs. *Conservation Biology* 18:639–45.

Wilcove, D. S., and L. L. Master. 2005. "How Many Endangered Species Are There in the United States?" *Frontiers in Ecology and the Environment* 3:414–20.

Williams, T. 1996. "Finding Safe Harbor." *Audubon*, January–February, 26–32.

Zhang, D. W. 2004. "Endangered Species and Timber Harvesting: The Case of Red-Cockaded Woodpeckers." *Economic Inquiry* 42:150–65.

Zhang, D., and S. R. Mehmood. 2002. "Safe Harbor for the Red-Cockaded Woodpecker: Private Forest Landowners Share Their Views." *Journal of Forestry* 100:24–29.

Chapter 4

Binns, J. A., P. M. Illgner, and E. L. Nel. 2001. "Water Shortage, Deforestation and Development: South Africa's Working for Water Programme." *Land Degradation and Development* 12:341–55.

Bond, W., and P. Slingsby. 1983. "Seeds Dispersal by Ants in Cape Shrublands and Its Evolutionary Implications." *South African Journal of Science* 79:231–33.

———. 1984. "Collapse of an Ant-Plant Mutalism: The Argentine Ant (*Iridomyrmex humilis*) and Myrmecochorous Proteaceae." *Ecology* 65:1031–37. doi:10.2307/1938311.

Cherry, M. I. 2005. "South Africa—Serious about Biodiversity Science." *PLoS Biology* 3:e145.

De Wit, M. P., D. J. Crookes, and B. W. van Wilgen. 2001. "Conflicts of Interest in Environmental Management: Estimating the Costs and Benefits of a Tree Invasion." *Biological Invasions* 3:167–78.

Dye, P., and C. Jarmain. 2004. "Water Use by Black Wattle (*Acacia mearnsii*): Implications for the Link between Removal of Invading Trees and Catchment Streamflow Response." *South African Journal of Science* 100:40–44.

Dye, P., G. Moses, P. Vilakazi, R. Ndlela, and M. Royappen. 2004. "Comparative Water Use of Wattle Thickets and Indigenous Plant Communities at Riparian Sites in the Western Cape and KwaZulu-Natal." *Water SA* 27:529–38.

Dye, P. J., and A. G. Poulter. 1995. "A Field Demonstration of the Effect on Streamflow of Clearing Invasive Pine and Wattle Trees from a Riparian Zone." *South African Forestry Journal* 173:27–30.

Gorgens, A. H. M., and B. W. van Wilgen. 2004. "Invasive Alien Plants and Water Resources in South Africa: Current Understanding, Predictive Ability and Research Challenges." *South African Journal of Science* 100:27–33.

Higgins, S. I., D. M. Richardson, R. M. Cowling, and T. H. Trinder Smith. 1999. "Predicting the Landscape Scale Distribution of Alien Plants and Their Threat to Plant Diversity." *Conservation Biology* 13:303–13.

Hosking, S. G., and M. Du Preez. 1999. "A Cost-Benefit Analysis of Removing Alien Trees in the Tsitsikamma Mountain Catchment." *South African Journal of Science* 95:442–48.

———. 2002. "Valuing Water Gains in the Eastern Cape's Working for Water Programme." *Water SA* 28:23–28.

———. 2004. "A Cost-Benefit Analysis of the Working for Water Programme on Selected Sites in South Africa." *Water SA* 30:143–52.

Koenig, R. 2009. "Unleashing an Army to Repair Alien-Ravaged Ecosystems." *Science* 325:562–63.

Lubke, R. A. 1985. "Erosion of the Beach at St. Francis Bay, Eastern Cape, South Africa." *Biological Conservation* 32:99–127.

Macdonald, I. A.W. 2004. "Recent Research on Alien Plant Invasions and Their Management in South Africa: A Review of the Inaugural Research Symposium of the Working for Water Programme." *South African Journal of Science* 100:21–26.

Magadlela, D., and N. Mdzeke. 2004. "Social Benefits in the Working for Water Programme as a Public Works Initiative." *South African Journal of Science* 100:94–96.

Le Maitre, D. C., B. W. van Wilgen, R. A. Chapman, and D. H. McKelly. 1996. "Invasive Plants and Water Resources in the Western Cape Province, South Africa: Modelling the Consequences of a Lack of Management." *Journal of Applied Ecology* 33:161–72.

Le Maitre, D. C., B. W. van Wilgen, C. M. Gelderblom, C. Bailey, R. A. Chapman, and J. A. Nel. 2002. "Invasive Alien Trees and Water Resources in South Africa: Case Studies of the Costs and Benefits of Management." *Forest Ecology and Management* 160:143–59.

Le Maitre, D. C., D. B. Versfeld, and R. A. Chapman. 2000. "The Impact of Invading Alien Plants on Surface Water Resources in South Africa: A Preliminary Assessment." *Water SA* 26:397–408.

Marais, C., B. W. van Wilgen, and D. Stevens. 2004. "The Clearing of Invasive Alien Plants in South Africa: A Preliminary Assessment of Costs and Progress." *South African Journal of Science* 100:97–103.

Mills, A. J., J. Turpie, R. M. Cowling, et al. 2007. "Assessing Costs, Benefits, and Feasibility of Subtropical Thicket in Restoration in the Eastern Cape, South Africa." In *Restoring Natural Capital: Science, Business and Practice*, edited by J Aronson, S. J. Milton, and J. N. Bilgnaut, 179–187. Washington DC: Society for Ecological Restoration, Island Press.

Prinsloo, F. W., and D. F. Scott. 1999. "Streamflow Responses to the Clearing of Alien Invasive Trees from Riparian Zones at Three Sites in the Western Cape Province." *Southern African Forestry Journal* 185:1–7.

Raimondo, D., and L. Von Staden. 2009. "Patterns and Trends in the Red List of South African Plants." *Strelitzia* 25:19–40.

Richardson, D. M., and B. W. van Wilgen. 2004. "Invasive Alien Plants in South Africa: How Well Do We Understand the Ecological Impacts?" *South African Journal of Science* 100:45–52.

Scott, D. F. 1999. "Managing Riparian Zone Vegetation to Sustain Streamflow: Results of Paired Catchment Experiments in South Africa." *Canadian Journal of Forestry Research* 29:1149–57.

Scott, D. F., L. A. Bruijnzeel, R. A. Vertessy, and I. R. Calder. 2004. "Forest Hydrology: Impacts of Forest Plantation on Streamflow." In *The Encyclopedia of Forest Sciences*, edited by J. Burnley, J. Evans, and J. A. Youngquist, 367–377. Oxford: Elsevier.

Scott, D. F., and R. E. Smith. 1997. "Preliminary Empirical Models to Predict Reductions in Total and Low Flows Resulting from Afforestation." *Water SA* 23:135–40.

Turpie, J. 2004. "The Role of Resource Economics in the Control of Invasive Alien Plants in South Africa." *South African Journal of Science* 100:87–93.

Van Wilgen, B. W. 1991. "Water Catchment Areas in the Fynbos: What of the Future?" *Veld and Flora*, June, 34–35.

———. 2004. "Scientific Challenges in the Field of Invasive Alien Plant Management." *South African Journal of Science* 100:19–20.

Van Wilgen, B. W., W. J. Bond, and D. M. Richardson. 1992. "Ecosystem Management." In *The Ecology of Fynbos: Nutrients, Fire and Diversity*, edited by R. M. Cowling, 345–71. Cape Town: Oxford University Press.

Van Wilgen, B. W., R. M. Cowling, and C. J. Burgers. 1996. "Valuation of Ecosystem Services." *BioScience* 46:184–89.

Van Wilgen, B. W., M. P. De Wit, H. J. Anderson, et al. 2004. "Costs and Benefits of Biological Control of Invasive Alien Plants: Case Studies from South Africa." *South African Journal of Science* 100:113–22.

Van Wilgen, B. W., D. C. Le Maitre, and R. M. Cowling. 1998. "Ecosystem Services, Efficiency, Sustainability and Equity: South Africa's Working for Water Programme. *Trends in Ecoloogy and Evolution* 13:378.

Van Wilgen, B. W., P. R. Little, R. A. Chapman, A. H. M. Gorgens, T. Willems, and C. Marais. 1997. "The Sustainable Development of Water Resources: History, Financial Costs, and Benefits of Alien Plant Control Programmes." *South African Journal of Science* 93:404–11.

Van Wilgen, B. W., C. Marais, D. Magadlela, N. Jezile, and D. Stevens. 2002. "Win—Win—Win: South Africa's Working for Water Programme." In *Mainstreaming Biodiversity in Development Case Studies from South Africa*, edited by S. M. Pierce, R. M. Cowling, T. Sandwith, and K. MacKinnon, 5–20. Washington DC: World Bank Environment Department.

Van Wilgen, B. W., D. M. Richardson, D. C. Le Maitre, C. Marais, and D. Magadlela. 2001. "The Economic Consequences of Alien Plant Invasions: Examples of Impacts and Approaches to Sustainable Management in South Africa." *Environment, Development and Sustainability* 3:145–68.

Van Wilgen, B. W., and E. Van Wyk. 1999. "Invading Alien Plants in South Africa: Impacts and Solutions." In *Proceedings of the VI International Rangeland Congress*, edited by D. Eldridge and D. Freudenberger, 566–71. Townsville, Australia: International Rangelands Congress Inc.

Woodworth, P. 2006. "Working for Water in South Africa: Saving the World on a Single Budget." *World Policy Journal* (Summer): 31–43.

World Resources Institute. 2000. "Working for Water, Working for Human Welfare in South Africa." In *A Guide to World Resources 2000–2001: People and Ecosystems: The Fraying Web of Life*, edited by World Resources Institute. Washington DC: World Resources Institute.

Zimmermann, H. G., V. C. Moran, and J. H. Hoffmann. 2004. "Biological Control in the Management of Invasive Alien Plants in South Africa, and the Role of the Working for Water Programme." *South African Journal of Science* 100:34–40.

Chapter 5

Beemster, N. 2001. "The Long-Term Influence of Grazing by Livestock on Vole-Feeding Raptors in the Netherlands." In *Hungry Herds: Management of Temperate Lowland Wetlands by Grazing*, unpublished master's thesis, edited by J. T. Vulink, 271–290. Lelystad: Van Zee tot Land.

Diamond, J. 1975. "The Island Dilemma: Lessons of Modern Biogeographic Studies for the Design of Natural Reserves." *Biological Conservation* 7:3–15.

Enserink, M., and G. Vogel. 2006. "The Carnivore Comeback." *Science* 314:746–49.

Fokkens, B. 2005. "The Dutch Strategy for Safety and River Flood Prevention". In *Extreme Hydrological Events: New Concepts for Security*, edited by O. F. Vasilier, P. H. A. J. M. Van Gelder, M. V. Bolgov, and F. J. Plate, 335–52. Berlin: Springer.

Green, R. E., S. J. Cornell, J. P. W. Scharlemann, and A. Balmford. 2005. "Farming and the Fate of Wild Nature." *Science* 307:550–55. doi:10.1126/science.1106049.

Hodder, K. H., and J. M. Bullock. 2005. "Nature without Nurture." *Planet Earth*, Winter, 30–31.

Hodder, K. H., J. M. Bullock, P. C. Buckland, and K. J. Kirby. 2005. "Large Herbivores in the Wildwood and Modern Naturalistic Grazing Systems." *English Nature Research Reports* no. 648. Peterborough, UK: English Nature.

Hoekstra, A., and P. Cornelissen. 2001. "Sexual Segregation in a Herd of Heck Cattle: The Occurrence of Bull Groups." In *Hungry Herds: Management of Temperate Lowland Wetlands by Grazing*, unpublished master's thesis, edited by J. T. Vulink, 105–27. Lelystad: Van Zee tot Land.

Hootsmans, M. A., and H. Kampf. 2004. *Ecological Networks: Experiences in the Netherlands*. The Hague: Ministry of Agriculture, Nature and Food Quality.

Huijser, M. P., P. Cornelissen, and M. Zijlstra. 2001. "Greylag Geese *Anser anser* Fattening Up in a Managed Wetland: Do Grazing Regimes Influence Habitat Selection?" In *Hungry Herds: Management of Temperate Lowland Wetlands by Grazing*, unpublished master's thesis, edited by J. T. Vulink, 221–237. Lelystad: Van Zee tot Land.

Kirby, K. J. 2004. "What Might a British Forest-Landscape Driven by Large Herbivores Look Like?" *Forestry* 77:405–20.

MacArthur, R. H., and E. O. Wilson. 1967. *The Theory of Island Biogeography*. Princeton, NJ: Princeton University Press.

Manning, A. D., I. J. Gordon, and W. J. Ripple. 2009. "Restoring Landscapes of Fear with Wolves in the Scottish Highlands." *Biological Conservation* 142:2314–21.

Mduma, S. A. R., A. R. E. Sinclair, and R. Hilborn. 1999. "Food Regulates the Serengeti Wildebeest: A 40 Year Record." *Journal of Animal Ecology* 68: 1101–22.

Ministry of Transport, Public Works and Water Management. 2006. *Spatial Planning Key Decision: "Room for the River": Investing in the Safety and Viability of the Dutch River Basin Region*. The Hague, Netherlands: Ministry of Transport, Public Works and Water Management.

Nolet, B. A. 2005. "Nature's Engineers: The Beavers' Return to the Netherlands." In *Seeking Nature's Limits: Ecologists in the Field*, edited by R. Drent, J. Tinbergers, J. P. Bakker, T. Piersma, G. P. Baerends, and S. J. Moore, 259–64. Zeist: KNNV Publishing.

Nolet, B. A., and J. M. Baveco. 1996. "Development and Viability of a Translocated Beaver *Castor fiber* Population in the Netherlands." *Biological Conservation* 75:125–37.

Nolet, B. A., and F. Rosell. 1998. "Comeback of the Beaver *Castor fiber*: An overview of Old and New Conservation Problems." *Biological Conservation* 83, no. 2: 165–73.

Pauly, D. 1995. "Anecdotes and the Shifting Baseline Syndrome of Fisheries." *Trends in Ecology and Evolution* 10:430.

Rosell, F., O. Bozsér, P. Collen, and H. Parker. 2005. "Ecological Impact of Beavers *Castor fiber* and *Castor canadensis* and Their Ability to Modify Ecosystems." *Mammal Review* 35:248–76.

Second International Commission on Management of the Oostvaardersplassen. 2010. *Natural Processes, Animal Welfare, Moral Aspects and Management of the Oostvaardersplassen. Report of the Second International Commission on Management of the Oostvaardersplassen (ICMO2)*. The Hague/Wageningen, the Netherlands: ICMO.

Theil, S. 2005. "Into the Woods." *Newsweek*, July 4, 24–28.

Vera, F. W. M. 1980. *The "Oostvaardersplassen": Possible Ways of Preserving and Further Developing of the Ecosystem*. The Netherlands: State Forest Service in the Netherlands.

———. 2000. *Grazing Ecology and Forest History*. Wallingford: CABI Publishing.

Voslamber, B., and J. T. Vulink. 2010. "Experimental Manipulation of Water Table and Grazing Pressure as a Tool for Developing and Maintaining Habitat Diversity for Waterbirds." *Ardea* 98:329–38.

Vulink, J. T., and M. R. Van Eerden. 1998. "Hydrological Conditions and Herbivores as Key Operators for Ecosystem Development in Dutch Artificial Wetlands." In *Grazing and Conservation Management*, edited by M. F. W. De Fries, J. P. Bakker, and S. E. Van Wieren, 217–52. Dordrecht: Kluwer.

Wigbels, V. 2001. *Oostvaardersplassen*. Zwolle, the Netherlands: Staatsbosbeheer.

World Wide Fund for Nature. 1999. *Living Rivers*. Zeist, the Netherlands: WWF.

Chapter 6

Altman, J. 2009. "*Manwurrk* (Fire Drive) at Namilewohwo: A Land Management, Hunting and Ceremonial Event in Western Arnhem Land." In *Culture, Ecology and Economy of Fire Management in North Australian Savannas: Rekindling the Wurrk Tradition.*, edited by J. Russell-Smith, P. Whitehead, and P. M. Cooke, 165–80. Collingwood, Australia: CSIRO.

Arriagada, R. A., S. K. Pattanayak, P. J. Ferraro, and E. O. Sills. 2009. "Combining Qualitative and Quantitative Methods to Evaluate Participation in Costa Rica's Program of Payments for Environmental Services." *Journal of Sustainable Forestry* 28:343–67.

Balmford, A., J. L. Moore, T. Brooks, et al. 2001. "Conservation Conflicts across Africa." *Science* 291:2616–19. doi:10.1126/science.291.5513.2616.

Becker, C. D. 1999. "Protecting a Garúa Forest in Ecuador: The Role of Institutions and Ecosystem Valuation." *Ambio* 28:156–61.

———. 2003. "Grassroots to Grassroots: Why Forest Preservation Was Rapid at Loma Alta, Ecuador." *World Development* 31:163–76.

Becker, C. D., A. Agreda, E. Astudillo, M. Costantino, and P. Torres. 2005. "Community Based Monitoring of Fog Capture and Biodiversity at Loma Alta, Ecuador Enhance Social Capital and Institutional Cooperation." *Biodiversity and Conservation* 14:2695–707.

Becker, C. D., and K. Ghimire. 2003. "Synergy between Traditional Ecological Knowledge and Conservation Science Supports Forest Preservation in Ecuador." *Conservation Ecology* 8, no. 1: 1.

Becker, C. D., and D. López-Lanús. 1997. "Conservation Value of a Garúa Forest in the Dry Season: A Bird Survey in the Reserva Ecológica de Loma Alta, Ecuador." *Cotinga* 8:66–74.

Bliege Bird, R., D. W. Bird, B. F. Codding, C. H. Parker, and J. H. Jones. 2008. The "Fire Stick Farming" Hypothesis: Australian Aboriginal Foraging Strategies, Biodiversity, and Anthropogenic Fire Mosaics." *Proceedings of the National Academy of Sciences* 105:14796–801. doi:10.1073/pnas.0804757105.

Bond, I. 2007. *Payments for Watershed Services: Opportunities and Realities.* London: International Institute for Environment and Development.

Bonifaz, C., and X. Cornejo. 2004. *Flora del Bosque de Garúa (árboles y epífitas) de la comuna Loma Alta, cordillera Chongón Colonche, provincia del Guayas, Ecuador.* Guayaquil, Ecuador: Universidad de Guayaquil, con la colaboración de Missouri Botanical Garden, Fundación Gaia.

Bruijnzeel, L. A. 2001. "Hydrology of Tropical Montane Cloud Forests: A Reassessment." *Land Use and Water Resources Research* 1:1.1–1.18.

———. 2005. "Tropical Montane Cloud Forest: A Unique Hydrological Case." In *Forests, Water and People in the Humid Tropics*, edited by M. Bonell and L. A. Bruijnzeel, 1:462–82. Cambridge: Cambridge University Press.

———. 2006. *Hydrological Impacts of Converting Tropical Montane Cloud Forest to Pasture, with Initial Reference to Northern Cost Rica.* Unpublished report. Amsterdam: Vrije Univesiteit, DFID.

Calder, I. R. 2004. "Forests and Water—Closing the Gap between Public and Science Perceptions." *Water Science and Technology* 49:39–53.

Chomitz, K. M., E. Brenes, and L. Constantino. 1999. "Financing Environmental Services: The Costa Rican Experience and Its Implications." *Science of the Total Environment* 240:157–69.

Chomitz, K. M., and K. Kumari. 1998. "The Domestic Benefits of Tropical Forests: A Critical Review." *World Bank Research Observer* 13, no. 1: 13–35.

Cooke, P. M. 2009. "Buffalo and Tin, Baki and Jesus: The Creation of a Modern Wilderness." In *Culture, Ecology and Economy of Fire Management in North Australian Savannas: Rekindling the Wurrk Tradition.*, edited by J. Russell-Smith, P. Whitehead, and P. M. Cooke, 69–83. Collingwood, Australia: CSIRO.

Daily, G. C., and K. Ellison. 2002. *The New Economy of Nature: The Quest to Make Nature Profitable.* Washington DC: Island Press/Shearwater Books.

Dodson, C. H., and A. H. Gentry. 1991. "Biological Extinction in Western Ecuador." *Annals of the Missouri Botanical Garden* 78:273–95.

Engel, S., S. Pagiola, and S. Wunder. 2008. "Designing Payments for Environmental Services in Theory and Practice: An Overview of the Issues." *Ecological Economics* 65:663–74.

Garde, M., B. L. Nadjamerrek, J. Kalarriya, et al. 2009. "The Language of Fire: Seasonality, Resources and Landscape Burning on the Arnhem Land Plateau." In

Culture, Ecology and Economy of Fire Management in North Australian Savannas. Rekindling the Wurrk Tradition., edited by J. Russell-Smith, P. Whitehead, and P. M. Cooke, 85–164. Collingwood, Australia: CSIRO.

Grubb, P. J. 1970. "Interpretation of the 'Massenerhebung' Effect on Tropical Mountains." *Nature* 229:44–45.

Hansen, M., R. De Fries, J. R. Townshend, M. Carroll, C. Dimiceli, and R. Sohlberg. "Vegetation Continuous Fields MOD44B, 2001 Percent Tree Cover, Collection 4." College Park, MD: University of Maryland.

Hardin, G. 1968. "The Tragedy of the Commons." *Science* 162:1243–48. doi:10.1126/science.162.3859.1243.

Hartshorn, G., P. J. Ferraro, and B. Spergel. 2005. *Evaluation of the World Bank—GEF Ecomarkets Project in Costa Rica*. Raleigh: North Carolina State University.

Ibarra, E. 2007. "The Profitability of Forest Protection versus Logging and the Role of Payments for Environmental Services (PES) in the Reserva Forestal Golfo Dulce, Costa Rica." *Forest Policy and Economics* 10:7–13.

Janzen, D. H. 1967. "Why Mountain Passes Are Higher in the Tropics." *American Naturalist* 101:233–47.

Jones, C. E. Jr., and P. V. Rich. 1972. "Ornithophily and Extrafloral Color Patterns in *Columnea florida* Morton (Gesneriaceae)." *Bulletin of the Southern California Academy of Sciences* 71, no. 3: 113–16.

Kosoy, N., M. Martinez-Tuna, R. Muradian, and J. Martinez-Alier. 2007. "Payments for Environmental Services in Watersheds: Insights from a Comparative Study of Three Cases in Central America." *Ecological Economics* 61:446–55.

Luck, G. W. 2007. A Review of the Relationships between Human Population Density and Biodiversity." *Biological Reviews* 82:607–45.

Meyer, S. 2010. "Working with Landowners to Provide Ecosystem Services: Costa Rica's Groundbreaking Experiment." In *Conservation Capital in the Americas: Exemplary Conservation Finance Initiatives in the Western Hemisphere.*, edited by J. N. Levitt, 207–219. Cambridge, MA: Lincoln Institute of Land Policy.

Miranda, M., I. T. Porras, and M. L. Moreno. 2003. *The Social Impacts of Payments for Environmental Services in Costa Rica*. London: International Institute for Environment and Development.

Ostrom, E. 1990. *Governing the Commons: The Evolution of Institutions for Collective Action*. Cambridge: Cambridge University Press.

Pagiola, S. 2002. "Paying for Water Services in Central America: Learning from Costa Rica." In *Selling Forest Environmental Services: Market-Based Mechanisms for Conservation and Development*, edited by S. Pagiola, J. Bishop, and N. Landel-Mills, 37–61. London: Earthscan.

———. 2008. "Payments for Environmental Services in Costa Rica." *Ecological Economics* 65, no. 4: 712–24.

Pagiola, S., P. Agostini, J. Gobbi, et al. 2004. *Paying for Biodiversity Conservation Services in Agricultural Landscapes: Environment Department Paper 96*. New York: World Bank.

Pagiola, S., A. Arcenas, and G. Platais. 2005. "Can Payments for Environmental Services Help Reduce Poverty? An Exploration of the Issues and the Evidence to Date from Latin America." *World Development* 33:237–53.

Pagiola, S., E. Ramírez, J. Gobbi, et al. 2007. "Paying for the Environmental Services of Silvopastoral Practices in Nícaragua." *Ecological Economics* 64:374–85.

Porras, I. T., and M. Grieg-Chan. 2007. *Watershed Services: Who Pays and For What?* London: International Institute for Environment and Development.

Ritchie, D. 2009. "Things Fall Apart: The End of an Era of Systematic Indigenous Fire Management." In *Culture, Ecology and Economy of Fire Management in North Australian Savannas: Rekindling the Wurrk Tradition.*, edited by J. Russell-Smith, P. Whitehead, and P. M. Cooke, 23–40. Collingwood, Australia: CSIRO.

Rosa, H., S. Kandel, and L. Dimas. 2004. "Compensation for Environmental Services and Rural Communities: Lessons from the Americas and Key Issues for Strengthening Community Strategies." *International Forestry Review* 6:187–94.

Sánchez-Azofeifa, G. A., A. Pfaff, J. A. Robalino, and J. P. Boomhower. 2007. "Costa Rica's Payment for Environmental Services Program: Intention, Implementation, and Impact." *Conservation Biology* 21:1165–73.

Sierra, R., and E. Russman. 2006. "On the Efficiency of Environmental Service Payments: A Forest Conservation Assessment in the Osa Peninsula, Costa Rica." *Ecological Economics* 59:131–41.

Sierra, R., and J. Stallings. 1998. "The Dynamics and Social Organization of Tropical Deforestation in Northwest Ecuador, 1983–1995." *Human Ecology* 26:135–61.

Sills, E. O., R. A. Arriagada, P. J. Ferraro, et al. 2009. "Impact of the PSA Program on Land Use." Chapter 9 in *Ecomarkets: Costa Rica's Experience with Payments for Environmental Services*, edited by G. Platais and S. Pagiola. Washington DC: World Bank.

Snider, A. G., S. K. Pattanayak, E. O. Sills, and J. L. Schuler. 2003. "Policy Innovations for Private Forest Management and Conservation in Costa Rica." *Journal of Forestry* 105, no. 5: 18–23.

Woinarski, J. C. Z., J. Russell-Smith, A. N. Anderson, and K. Brennan. 2009. "Fire Management and Biodiversity of the Western Arnhem Land Plateau." In *Culture, Ecology and Economy of Fire Management in North Australian Savannas. Rekindling the Wurrk Tradition.*, edited by J. Russell-Smith, P. Whitehead, and P. M. Cooke, 201–27. Collingwood, Australia: CSIRO.

Wunder, S. 2007. "The Efficiency of Payments for Environmental Services in Tropical Conservation." *Conservation Biology* 21:48–58.

Yates, C. P., A. C. Edwards, and J. Russell-Smith. 2008. "Big Fires and Their Ecological Impacts in Australian Savannas: Size and Frequency Matters." *International Journal of Wildland Fire* 17:768–81.

Zbinden, S., and D. R. Lee. 2005. "Paying for Environmental Services: An Analysis of Participation in Costa Rica's PSA Program." *World Development* 33:255–72.

Chapter 7

Abbott, I. 2002. "Origin and Spread of the Cat, *Felis catus*, on Mainland Australia, with a Discussion of the Magnitude of Its Early Impact on Native Fauna." *Wildlife Research* 29: 51–74.

Alcoa World Alumina Australia. 2003. *Restoring the Botanical Richness of the Jarrah Forest after Bauxite Mining in South-Western Australia*. Booragoon, Australia: Alcoa World Alumina Australia.

———. 2006. *Mining Environmental Improvement Plan 2006–2007*. Booragoon, Australia: Alcoa World Alumina Australia.

Algar, D., and N. D. Burrows. 2004. "Feral Cat Control Research. Western Shield Review—February 2003." *Conservation Science Western Australia* 5:131–63.

Armstrong, R. 2004. "Baiting Operations: Western Shield Review —February 2003." *Conservation Science Western Australia* 5:31–50.

Bailey, C. 1996. "Western Shield: Bringing Wildlife Back from the Brink of Extinction." *Landscope* 11:41–48.

Bartle, J., and G. C. Slessar. 1989. "The Jarrah Forest, an Introduction." In *The Jarrah Forest: A Complex Mediterranean Ecosystem.*, edited by B. Dell, J. J. Havel, and N. Malajezuk, 357–77. Dordrecht: Kluwer.

Bell, D. T., and R. J. Mead. 2007. "Jarrah Forest Ecosystem Restoration: A Foreword." *Restoration Ecology* 15:S1–S2.

Blyth, J., and A. A. Burbidge. 2004. "Threatened Fauna Issues Not Covered under Western Shield: Western Shield Review—February 2003." *Conservation Science Western Australia* 5:164–73.

Brennan, K. E. C., O. G. Nichols, and J. D. Majer. 2005. *Innovative Techniques for Promoting Fauna Return to Rehabilitated Sites following Mining*. Brisbane: Australian Centre for Minerals Extraction and Research, Minerals and Energy Research Institute of Western Australia.

Colquhoun, I. J., and N. L. Kerp. 2007. "Minimizing the Spread of a Soil Borne Plant Pathogen during a Large Scale Mining Operation." *Restoration Ecology* 15:S85–S93.

Croton, J. T., and G. L. Ainsworth. 2007. "Development of a Winged Tine to Relieve Mining Related Soil Compaction after Bauxite Mining in Western Australia." *Restoration Ecology* 15:S48–S53.

Dell, B., and J. J. Havel. 1989. "The Jarrah Forest, an Introduction." In *The Jarrah Forest: A Complex Mediterranean Ecosystem.*, edited by B. Dell, J. J. Havel, and N. Malajezuk, 1–10. Dordrecht: Kluwer.

Dixon, K. W., S. Roche, and J. S. Pate. 1995. "The Promotive Effect of Smoke Derived from Burnt Native Vegetation on Seed Germination of Western Australian Plants." *Oecologia* 101:185–92.

Flematti, G. R., E. L. Ghisalberti, K. W. Dixon, and R. D. Trengove. 2004. "A Compound from Smoke that Promotes Seed Germination." *Science* 305:977.

Friend, J. A. 1990. "The Numbat *Myrmecobius fasciatus* (Myrmecobiidae): History of Decline and Potential for Recovery." *Proceedings of the Ecological Society of Australia* 16:369–77.

Friend, T., and B. Beecham. 2004. "Return to Dryandra: Western Shield Review—February 2003." *Conservation Science Western Australia* 5:174–93.

Gardner, J. 2001. "Rehabilitating Mines to Meet Land Use Objectives: Bauxite Mining in the Jarrah Forest of Western Australia." *Unasylva (FAO)* 207, no. 52: 3–8.

Gardner, J. H., and D. T. Bell. 2007. "Bauxite Mining Restoration by Alcoa World Alumina Australia in Western Australia: Social, Political, Historical, and Environmental Contexts." *Restoration Ecology* 15:S3–S10.

Gole, C. 2006. *The Southwest Australia Ecoregion: Jewel of the Australian Continent.* Wembley, Western Australia: Southwest Australia Ecoregion Initiative.

Grant, C. D., S. C. Ward, and S. C. Morley. 2007. "Return of Ecosystem Function to Restored Bauxite Mines in Western Australia." *Restoration Ecology* 15:S94–S103.

Grant, C., and J. Gardner. 2005. "Mainstreaming Biodiversity in the Mining Industry: Experiences from Alcoa's Bauxite Mining Operations in Western Australia." In *Mainstreaming Biodiversity in Production Landscapes*, edited by C. Petersen and B. Huntley, 142–153. Washington DC: Global Environment Facility.

Johnson, C. 2006. *Australia's Mammal Extinctions.* Cambridge: Cambridge University Press.

Johnson, C. N., J. L. Isaac, and D. O. Fisher. 2007. "Rarity of a Top Predator Triggers Continent-Wide Collapse of Mammal Prey: Dingoes and Marsupials in Australia." *Proceedings of the Royal Society B: Biological Sciences* 274:341–46.

King, D. R., A. J. Oliver, and R. J. Mead. 1978. "The Adaptation of Some Western Australian Mammals to Food Plants Containing Fluoroacetate." *Australian Journal of Zoology* 26:699–712.

Kinnear, J. E., M. L. Onus, and N. R Sumner. 1998. "Fox Control and Rock-Wallaby Population Dynamics II: An Update." *Wildlife Research* 25:81–88.

Koch, J. M. 2006. "Alcoa's Mine Restoration in Western Australia: Philosophies, Structures and Strategies for Continued Improvement." Paper presented at the meeting of the Minerals Council of America. http:// www.minerals.org.au/ _data/assets/pdf_file/0015/17070/JohnKoch.pdf.

———. 2007a. "Restoring a Jarrah Forest Understorey Vegetation after Bauxite Mining in Western Australia." *Restoration Ecology* 15:S26–S39.

———. 2007b. "Alcoa's Mining and Restoration Process in South Western Australia. *Restoration Ecology* 15:S11–S16.

Koch, J. M., and G. P. Samsa. 2007. "Restoring Jarrah Forest Trees after Bauxite Mining in Western Australia." *Restoration Ecology* 15:S17–S25.

Majer, J. D., K. E. C. Brennan, and M. L. Moir. 2007. "Invertebrates and the Restoration of a Forest Ecosystem: 30 Years of Research following Bauxite Mining in Western Australia." *Restoration Ecology* 15:S104–S115.

Marchesan, D., and S. M. Carthew. 2004. "Autecology of the Yellow-Footed Antechinus (*Antechinus flavipes*) in a Fragmented Landscape in Southern Australia." *Wildlife Research* 31, no. 3: 273–82.

Mawson, P. R. 2004. "Translocations and Fauna Reconstruction Sites: Western Shield Review—February 2003." *Conservation Science Western Australia* 5:108–21.

McIlroy, J. C. 1982. "The Sensitivity of Australian Animals to 1080 poison III: Marsupial and Eutherian Herbivores." *Wildlife Research* 9:487–503.

McNamara, K. 2004. "Foreword: Western Shield Review—February 2003." *Conservation Science Western Australia* 5:1.

Miranda, M., P. Burris, J. F. Bingcang, et al. 2003. *Mining and Critical Ecosystems: Mapping the Risks*. Washington DC: WRI.

Mullins, R. G., J. M. Koch, and S. C. Ward. 2002. "Practical Method of Germination for a Key Jarrah Forest Species: Snottygobble (*Persoonia longifolia*)." *Ecological Management and Restoration* 3:97–103.

Nichols, O. G., and C. Grant. 2007. "Vertebrate Fauna Recolonization of Restored Bauxite Mines—Key Findings from almost 30 Years of Monitoring and Research." *Restoration Ecology* 15:S116–S126.

Orell, P. 2004. "Fauna Monitoring and Staff Training: Western Shield Review—February 2003." *Conservation Science Western Australia* 5:51–91.

Possingham, H., P. Jarman, and A. Kearns. 2004. "Independent Review of Western Shield—February 2003." *Conservation Science Western Australia* 5:2–18.

Richards, D., and A. Parsons. 2004. "International Council on Mining and Metals Perspective on the IUCN Protected Areas Management Category System." *Parks* 14:39–45.

Roche, S., J. M. Koch, and K. W. Dixon. 1997. "Smoke Enhanced Seed Germination for Mine Rehabilitation in the Southwest of Western Australia." *Restoration Ecology* 5:191–203.

Rossetto, M., F. Lucarotti, S. D. Hopper, and K. W. Dixon. 1997. "DNA Fingerprinting of *Eucalyptus graniticola*: A Critically Endangered Relict Species or a Rare Hybrid?" *Heredity* 79:310–18.

Shearer, B. L., C. E. Crane, and A. Cochrane. 2004. "Quantification of the Susceptibility of the Native Flora of the South-West Botanical Province, Western Australia, to *Phytophthora cinnamomi*." *Australian Journal of Botany* 52:435–43.

Stolton, S., N. Dudley, and P. Toyne. 2001. *The UK's Forest Footprint*. Godalming, UK: WWF-UK.

Willyams, D. 2005. "Tissue Culture of Geophytic Rush and Sedge Species for Revegetation of Bauxite Mine Sites in the Northern Jarrah Forest of Western Australia." In *Proceedings of the Australian Branch of the International Association for Plant Tissue Culture and Biotechnology*, 226–241. Perth, Australia: International Association for Plant Tissue Culture and Biotechnology.

World Conservation Union (IUCN), and International Council on Minerals and Mining. 2004. *Integrating Mining and Biodiversity Conservation: Case Studies from around the World*. Gland, Switzerland: IUCN.

WWF Australia, and Dieback Consultative Coucil. 2004. *Arresting Phytophthora Dieback: The Biological Bulldozer*. Sydney: WWF Australia.

Wyre, G. 2004. "Management of the Western Shield Program: Western Shield Review—February 2003." *Conservation Science Western Australia* 5:20–30.

Chapter 8

Agnew, D. 2008. "Case Study 1: Toothfish an MSC Certified Fishery." In *Seafood Ecolabelling: Principles and Practice*, edited by T. J. Ward and B. Phillips, 247–58. Oxford: Wiley-Blackwell.

Agnew, D. J., C. Grieve, P. Orr, G. Parkes, and N. Barker. 2006. *Environmental Benefits Resulting from Certification against MSC's Principles and Criteria*. London: MRAG UK, Marine Stewardship Council.

Agnew, D. J., J. Pearce, G. Pramod, et al. 2009. Estimating the Worldwide Extent of Illegal Fishing. *PLoS One* 4:e4570.

Arnason, R., K. Kelleher, and R. Willmann. 2009. *The Sunken Billions: The Economic Justification for Fisheries Reform*. Washington DC, Rome: World Bank, FAO.

Baum, J. K., and R. A. Myers. 2004. "Shifting Baselines and the Decline of Pelagic Sharks in the Gulf of Mexico." *Ecology Letters* 7:135–45.

Beddington, J. R., D. J. Agnew, and C. W. Clark. 2007. "Current Problems in the Management of Marine Fisheries." *Science* 316:1713–16.

Branch, T. A., R. Watson, E. A. Fulton, et al. 2010. "The Trophic Fingerprint of Marine Fisheries." *Nature* 468:431–35.

Cambridge, T., S. Martin, F. Nimmo, et al. 2011. *Researching the Environmental Impacts of the MSC Certification Programme*. London: MRAG UK.

Christensen, V., P. Amorim, I. Diallo, et al. 2004. "Trends in Fish Biomass Off Northwest Africa, 1960–2000." In *West African Marine Ecosystems: Models and Fisheries Impacts*, edited by M. L. D. Palomares and D. Pauly, 215–20. Vancouver: Fisheries Centre University of British Columbia.

Christensen, V., S. Guénette, J. J. Britcher, et al. 2003. "Hundred Year Decline of North Atlantic Predatory Fishes." *Fish and Fisheries* 4:1–24.

Clover, C. 2005. *The End of the Line: How Overfishing Is Changing the World and What We Eat*. London: Ebury Press.

Collie, J. S., S. J. Hall, M. J. Kaiser, and I. R. Poiner. 2000. "A Quantitative Analysis of Fishing Impacts on Shelf Sea Benthos." *Journal of Animal Ecology* 69: 785–98.

Cooper, L. 1996. "Don't Be Harsh on the MSC." *Samudra*, July, 12.

Costello, C., S. D. Gaines, and J. Lynham. 2008. "Can Catch Shares Prevent Fisheries Collapse?" *Science* 321:1678–81.

Davies, R. W. D., and R. Rangeley. 2010. "Banking on Cod: Exploring Economic Incentives for Recovering Grand Banks and North Sea Cod Fisheries." *Marine Policy* 34:92–98.

Ellis, R. 2008. "The Bluefin in Peril." *Scientific American* 298, no. 3: 58–65.

Food and Agriculture Organization of the United Nations. 2010. *The State of World Fisheries and Aquaculture 2010*. Rome: FAO.

Gilmore, J. 2008. "Case Study 3: MSC Certification of the Alaska Pollock Fishery." In *Seafood Ecolabelling: Principles and Practice*, edited by T. J. Ward and B. Phillips, 269–86. Oxford: Wiley-Blackwell.

Gulbrandsen, L. H. 2009. "The Emergence and Effectiveness of the Marine Stewardship Council." *Marine Policy* 33:654–60.

Gutiérrez, N. L., R. Hilborn, and O. Defeo. 2011. "Leadership, Social Capital and Incentives Promote Successful Fisheries." *Nature* 470:386–89.

Hall, M., D. Pauly, D. Conover, et al. 2007. "10 Solutions to Save the Ocean." *Conservation Magazine* 8, no. 3: 23.

Hawken, P., and L. Niznik. 1993. *The Ecology of Commerce*. London: Harper Collins.

Hilborn, R., and J. H. Cowan Jr. 2010. "Marine Stewardship: High Bar for Seafood." *Nature* 467:531.

Hoel, A. H. 2004. *Ecolabelling in Fisheries: An Effective Conservation Tool?* Unpublished report. Tromsø, Norway: Norut Samfunnsforskning.

Hough, A., and P. Knapman. 2010. "Marine Stewardship: Fair and Impartial." *Nature* 467:531.

Howes, R. 2008. "The Marine Stewardship Council Programme." In *Seafood Ecolabelling: Principles and Practice*, edited by T. J. Ward and B. Phillips, 81–105. Oxford: Wiley-Blackwell.

———. 2010. "Marine Stewardship: Catalysing Change." *Nature* 467:1047.

Jackson, J. B. C. 2008. "Ecological Extinction and Evolution in the Brave New Ocean." *Proceedings of the National Academy of Sciences* 105:11458–65.

Jacquet, J. L., and D. Pauly. 2007. "The Rise of Seafood Awareness Campaigns in an Era of Collapsing Fisheries." *Marine Policy* 31:308–13.

———. 2008a. "Funding Priorities: Big Barriers to Small Scale Fisheries." *Conservation Biology* 22:832–35.

———. 2008b. "Trade Secrets: Renaming and Mislabeling of Seafood." *Marine Policy* 32:309–18.

Jacquet, J. L., J. Hocevar, S. Lai, et al. 2010. "Conserving Wild Fish in a Sea of Market-Based Efforts." *Oryx* 44:45–56.

Jacquet, J. L., D. Pauly, D. Ainley, S. Holt, P. Dayton, and J. Jackson. 2010. "Seafood Stewardship in Crisis." *Nature* 467:28–29.

Jennings, S., and J. L Blanchard. 2004. "Fish Abundance with no Fishing: Predictions Based on Macroecological Theory." *Journal of Animal Ecology* 73:632–42.

Kaiser, M. J., and G. Edwards-Jones. 2006. "The Role of Ecolabeling in Fisheries Management and Conservation." *Conservation Biology* 20:392–98.

Kaiser, M. J, and L. Hill. 2010. "Marine Stewardship: A Force for Good." *Nature* 467:531.

Leadbitter, D., G. Gomez, and F. McGilvray. 2006. "Sustainable Fisheries and the East Asian Seas: Can the Private Sector Play a Role?" *Ocean and Coastal Management* 49:662–75.

Lopuch, M. 2008. "Benefits of Certification for Small-Scale Fisheries." In *Seafood Ecolabelling: Principles and Practice*, edited by T. J. Ward and B. Phillips, 307–21. Oxford: Wiley-Blackwell.

Lotze, H. K., and B. Worm. 2009. "Historical Baselines for Large Marine Animals." *Trends in Ecology and Evolution* 24:254–62.

Majkowski, J. 2007. *Global Fishery Resources of Tuna and Tuna-Like Species*. Rome: FAO.

Matthew, S. 1998. "When Sandals Meet Suits." *Samudra*, January, 31–35.

May, B., D. Leadbitter, M. Sutton, and M. Weber. 2003. "The Marine Stewardship Council (MSC)." In *Eco-Labelling in Fisheries, What Is It All About?*, edited by B. Phillips, T. J. Ward, and C. Chaffee, 14–33. Oxford: Blackwell.

Ménard, F., A. Fonteneau, D. Gaertner, V. Nordstrom, B. Stéquert, and E. Marchal. 2000. "Exploitation of Small Tunas by a Purse-Seine Fishery with Fish Aggregating Devices and Their Feeding Ecology in an Eastern Tropical Atlantic Ecosystem." *ICES Journal of Marine Science* 57:525–30.

Myers, R. A., J. K. Baum, T. D. Shepherd, S. P. Powers, and C. H. Peterson. 2007. "Cascading Effects of the Loss of Apex Predatory Sharks from a Coastal Ocean." *Science* 315:1846–50.

O'Riordan, B. 1996. "Who's Being Seduced?" *Samudra*, July, 10–11.

Parkes, G., S. Walmsley, T. Cambridge, et al. 2009. *Review of Fish Sustainability Information Schemes—Final Report*. London: MRAG.

Penteriani, V., N. Pettorelli, I. J. Gordon, et al. 2010. "New European Union Fisheries Regulations Could Benefit Conservation of Marine Animals." *Animal Conservation* 13:1–2.

Pitcher, T., D. Kalikoski, G. Pramod, and K. Short. 2009. "Not Honouring the Code." *Nature* 457:658–59.

Ponte, S. 2008. "Greener than Thou: The Political Economy of Fish Ecolabeling and Its Local Manifestations in South Africa." *World Development* 36:159–75.

Purvis, A. 2009. *Net Benefits, The First Ten Years of MSC Certified Sustainable Fisheries*. London: MSC.

Roberts, C. 2007. *The Unnatural History of the Sea: The Past and Future of Humanity and Fishing*. London: Gaia.

Roheim, C. A., F. Asche, and J. I. Santos. 2011. "The Elusive Price Premium for Eco-labelled Products: Evidence from Seafood in the UK Market." *Journal of Agricultural Economics*. doi:10.1111/j.1477-9552.2011.00299.x.

Romanov, E. V. 2002. "Bycatch in the Tuna Purse-Seine Fisheries of the Western Indian Ocean." *Fishery Bulletin* 100:90–105.

———. 2008. "Bycatch and Discards in the Soviet Purse Seine Tuna Fisheries on FAD-Associated Schools in the North Equatorial Area of the Western Indian Ocean." *Journal of Marine Science* 7:163–74.

Rose, G. A. 2004. "Reconciling Overfishing and Climate Change with Stock Dynamics of Atlantic Cod (*Gadus morhua*) over 500 Years." *Canadian Journal of Fisheries and Aquatic Sciences* 61:1553–57.

Rosenberg, A. A., W. J. Bolster, K. E. Alexander, W. B. L. Leavenworth, A. B. Cooper, and M. G. McKenzie. 2005. "The History of Ocean Resources: Modeling Cod Biomass Using Historical Records." *Frontiers in Ecology and the Environment* 3:84–90.

Rosenberg, A. A., J. H. Swasey, and M. Bowman. 2006. "Rebuilding US Fisheries: Progress and Problems." *Frontiers in Ecology and the Environment* 4:303–8.

Safina, C., and D. H. Klinger. 2008. "Collapse of Bluefin Tuna in the Western Atlantic." *Conservation Biology* 22:243–46.

Schmidt, C. C. 1998. "Open and Transparent." *Samudra*, January, 24.

Sibert, J., J. Hampton, P. Kleiber, and M. Maunder. 2006. "Biomass, Size, and Trophic Status of Top Predators in the Pacific Ocean." *Science* 314:1773–76.

Smith, M. D., C. A. Roheim, L. B. Crowder, et al. 2010. "Sustainability and Global Seafood." *Science* 327:784–86.

Stokstad, E. 2010. "Behind the Eco-Label, a Debate Over Antarctic Toothfish." *Science* 329:1596.

Stokstad, E. 2011. "Seafood Eco-Label Grapples with Challenge of Proving Its Impact." *Science* 334:746.

Sumaila, U. R., and D. Pauly. 2007. "All Fishing Nations Must Unite to Cut Subsidies." *Nature* 450:945.

Sutton, D. 2003. "An Unsatisfactory Encounter with the MSC: A Conservation Perspective." In *Eco-Labelling in Fisheries, What Is It All About?* edited by B. Phillips, T. J. Ward, and C. Chaffee, 114–19. Oxford: Blackwell.

Sutton, M. 1998. "An Appeal for Co-operation." *Samudra*, January, 26–30.

Sutton, M., and L. Wimpee. 2008. "Towards Sustainable Seafood: The Evolution of a Conservation Movement." In *Seafood Ecolabelling: Principles and Practice*, edited by T. J. Ward and B. Phillips, 403–15. Oxford: Wiley-Blackwell.

Thurstan, R. H., S. Brockington, and C. M. Roberts. 2010. "The Effects of 118 Years of Industrial Fishing on UK Bottom Trawl Fisheries." *Nature Communications* 1, no. 15: 1–6. doi: 10.1038/ncomms1013.

Ward, P., and R. A. Myers. 2005. "Shifts in Open-Ocean Fish Communities Coinciding with the Commencement of Commercial Fishing." *Ecology* 86:835–47.

Ward, T. J. 2008. "Barriers to Biodiversity Conservation in Marine Fishery Certification." *Fish and Fisheries* 9:169–77.

Ward, T. J., and B. Phillips. 2008a. "Ecolabelling of Seafood: The Basic Concepts." In *Seafood Ecolabelling: Principles and Practice*, edited by T. J. Ward and B. Phillips, 1–38. Oxford: Wiley-Blackwell.

———. 2008b. "Anecdotes and Lessons of a Decade." In *Seafood Ecolabelling: Principles and Practice*, edited by T. J. Ward and B. Phillips, 416–35. Oxford: Wiley-Blackwell.

Watson, R., and D. Pauly. 2001. "Systematic Distortions in World Fisheries Catch Trends." *Nature* 414:534–36. doi:10.1038/35107050.

Worm, B., E. B. Barbier, N. Beaumont, et al. 2006. "Impacts of Biodiversity Loss on Ocean Ecosystem Services." *Science* 314:787–90.

Worm, B., R. Hilborn, J. K. Baum, et al. 2009. "Rebuilding Global Fisheries." *Science* 325:578–85.

Zeller, D., and D. Pauly. 2005. "Good News, Bad News: Global Fisheries Discards Are Declining, but so Are Total Catches." *Fish and Fisheries* 6:156–59.

Zolezzi, J. H., and L. D. Bradley Jr. 2008. "The Story of the San Diego Tuna Fleet." *Mains'l Haul* 44, no. 1–2:8–27.

Chapter 9

Balmford, A. 2009. "Saving Ely's Wildspace." *Cambridgeshire Bird Club Bulletin* 405:5–6.

Balmford, A., K. J. Gaston, S. Blyth, A. James, and V. Kapos. 2003. "Global Variation in Terrestrial Conservation Costs, Conservation Benefits, and Unmet Conservation Needs." *Proceedings of the National Academy of Sciences* 100:1046–50. doi:10.1073/pnas.0236945100.

Balmford, A., and T. Whitten. 2003. "Who Should Pay for Tropical Conservation, and How Could the Costs Be Met?" *Oryx* 37:238–50.

BBC. 2010. "Amazon Deforestation 'Down Again.'" *BBC.* http://www.bbc.co.uk/news/world-latin-america-11888875.

———. 2011. "Huge Rise in Brazil Deforestation." *BBC.* http://www.bbc.co.uk/news/world-latin-america-13449792.

Benning, T. L., D. Agnew, C. T. Atkinson, and P. M. Vitousek. 2002. "Interactions of Climate Change with Biological Invasions and Land Use in the Hawaiian Islands: Modeling the Fate of Endemic Birds Using a Geographic Information System." *Proceedings of the National Academy of Sciences* 99:14246–14249. doi:10.1073/pnas.162372399.

BirdLife International. 2010. "Species Factsheet: *Nipponia nippon.*" *BirdLife.* http://www.birdlife.org/datazone/speciesfactsheet.php?id=3801.

Bruner, A. G., R. E. Gullison, and A. Balmford. 2004. "Financial Costs and Shortfalls of Managing and Expanding Protected-Area Systems in Developing Countries." *BioScience* 54:1119–26.

Constable, A. J., W. K. de la Mare, D. J. Agnew, I. Everson, and D. Miller. 2000. "Managing Fisheries to Conserve the Antarctic Marine Ecosystem: Practical Implementation of the Convention on the Conservation of Antarctic Marine Living Resources (CCAMLR)." *ICES Journal of Marine Science* 57:778–91.

Croxall, J. P., and S. Nicol. 2004. "Management of Southern Ocean Fisheries: Global Forces and Future Sustainability." *Antarctic Science* 16:569–84.

Freed, L. A., R. L. Cann, M. L. Goff, W. A. Kuntz, and G. R. Bodner. 2005. "Increase in Avian Malaria at Upper Elevation in Hawai'i." *Condor* 107:753–64.

Green, R. E., W. G. Hunt, C. N. Parish, and I. Newton. 2008. "Effectiveness of Action to Reduce Exposure of Free-Ranging California Condors in Arizona and Utah to Lead from Spent Ammunition." *PLoS One* 3: e4022. doi:10.1371/journal.pone.0004022.

Gross, L. 2005. "As the Antarctic Ice Pack Recedes, a Fragile Ecosystem Hangs in the Balance." *PLoS Biology* 3:e127.

Jackson, J. B. C., M. X. Kirby, W. H. Berger, et al. 2001. "Historical Overfishing and the Recent Collapse of Coastal Ecosystems." *Science* 293:629–37. doi:10.1126/science.1059199.

James, A. N., K. J. Gaston, and A. Balmford. 1999. "Balancing the Earth's Accounts." *Nature* 401:323–24. doi:10.1038/43774.

———. 2001. "Can We Afford to Conserve Biodiversity?" *BioScience* 51:43–52.

Leader-Williams, N., and S. D. Albon. 1988. "Allocation of Resources for Conservation." *Nature* 336:533–35.

Liu, J., and J. Diamond. 2008. "Revolutionizing China's Environmental Protection." *Science* 319:37–38. doi:10.1126/science.1150416.

Nelson, A., and K. M. Chomitz. 2009. *Protected Area Effectiveness in Reducing Tropical Deforestation: A Global Analysis of the Impact of Protection Status.* Washington DC: World Bank.

Nepstad, D. C., S. Schwartzman, B. Bamberger, et al. 2006. "Inhibition of Amazon Deforestation and Fire by Parks and Indigenous Lands." *Conservation Biology* 20:65–73.

Nepstad, D., B. S. Soares-Filho, F. Merry, et al. 2009. "The End of Deforestation in the Brazilian Amazon." *Science* 326:1350–51.

Nepstad, D. C., C. M. Stickler, and O. T. Almeida. 2006. "Globalization of the Amazon Soy and Beef Industries: Opportunities for Conservation." *Conservation Biology* 20:1595–603.

Nepstad, D. C., C. M. Stickler, B. S. Soares-Filho, and F. Merry. 2008. "Interactions among Amazon Land Use, Forests and Climate: Prospects for a Near-Term Forest Tipping Point." *Philosophical Transactions of the Royal Society of London B: Biological Sciences* 363:1737–46.

Parish, C. 2008. "A Hunter's New Perspective on Lead—Now He Shoots Solid Copper Bullets!" *Peregrine Fund Newsletter* 39 (Fall): 6–7.

Pew Research Center. 2008. *The 2008 Pew Global Attitudes Survey in China: The Chinese Celebrate Their Roaring Economy, as They Struggle with Its Costs.* Washington DC: Pew Research Center.

Platt, J. 2009. "Fight to Protect California Condors from Lead Ammunition Moves to Arizona." *Scientific American.* http://www.scientificamerican.com/blog/post .cfm?id=fight-to-protect-california-condors-2009-11-20.

Regalado, A. 2010. "Brazil Says Rate of Deforestation in Amazon Continues to Plunge." *Science* 329:1270–71.

Secretariat of the Convention on Biological Diversity. 2010. *Global Biodiversity Outlook 3.* Montreal: Secretariat of the Convention on Biological Diversity.

Sieg, R., K. A. Sullivan, and C. N. Parish. 2009. "Voluntary Lead Reduction Efforts within the Northern Arizona Range of the California Condor." In *Ingestion of Lead from Spent Ammunition: Implications for Wildlife and Humans,* edited by R. T. Watson, M. Fuller, M. Pokras, and G. Hunt, 341–49. Boise, ID: Peregrine Fund.

Small, C. J. 2005. *Regional Fisheries Management Organisations: Their Duties and Performance in Reducing Bycatch of Albatrosses and Other Species.* Cambridge: BirdLife International.

Steffen, W., P. J. Crutzen, and J. R. McNeill. 2007. "The Anthropocene: Are Humans Now Overwhelming the Great Forces of Nature?" *Ambio* 36:614–21.

Vreugdenhil, D. 2003. "Modelling the Financial Needs of Protected Area Systems: An Application of the 'Minimum Conservation System' Design Tool." Paper presented at the Fifth World Parks Congress, Durban, South Africa, September 8–17.

Xiaofeng, G. 2005. "Farmer the Biggest Hero Yew Can Find." *China Daily*. http://www.chinadaily.com.cn/english/doc/2005-05/20/content_444237.htm.

Zhan, X., M. Li, Z. Zhang, et al. 2006. "Molecular Censusing Doubles Giant Panda Population Estimate in a Key Nature Reserve." *Current Biology* 16:R451–R452.

Appendix

Barnes, S. 2007. *How to be Wild*. London: Short Books.

Bird, W. 2004. *Natural Fit. Can Green Space and Biodversity Increases Levels of Physical Activity?* Sandy, UK: Royal Society for the Protection of Birds.

Bird, W. 2007. *Natural Thinking. Investigating the Links between the Natural Environment, Biodiversity and Mental Health*. Sandy, UK: Royal Society for the Protection of Birds

Fairlie, S. 2010. *Meat. A Benign Extravagance*. East Meon, UK: Permanent Publications.

Gill, T. 2007. *Growing up in a Risk Averse Society*. London: Calouste Gulbenkian Foundation.

Jackson, T. 2009. *Prosperity without Growth: Economics for a Finite Planet*. London: Earthscan.

Keller, H. 1903. *Optimism: An Essay*. New York: T. Y. Crowell and Company.

Leopold, A. 1949. *A Sand County Almanac and Sketches Here and There*. New York: Oxford University Press.

Louv, R. 2007. *Last Child in the Woods. Saving Our Children from Nature Deficit Disorder*. Chapel Hill, NC: Algonquin Books.

MacKay, D. 2009. *Sustainable Energy—Without the Hot Air*. Cambridge: UIT.

Nabhan, G. P., and S. Trimble. 1995. *The Geography of Childhood: Why Children Need Wild Places*. Boston: Beacon Press.

Pelletier, N., and P. Tyedmers. 2010. "Forecasting Potential Global Environmental Costs of Livestock Production 2000–2050." *Proceedings of the National Academy of Sciences* 107:18371–74.

Peters, G. P., J. C. Minx, C. L. Weber, and O. Edenhofer. 2011. "Growth in Emissions Transfers via International Trade from 1990 to 2008." *Proceedings of the National Academy of Sciences* 108:8903–8.

Pyle, R. M. 2003. "Nature Matrix: Reconnecting People and Nature." *Oryx* 37: 206–14.

Raimondo, D., and L. Von Staden. 2009. "Patterns and Trends in the Red List of South African Plants." *Strelitzia* 25:19–40.

Stokstad, E. 2010. "Could Less Meat Mean More Food?" *Science* 327:810–11.

Weber, C. L., and H. S. Matthews. 2008. "Food-Miles and the Relative Climate Impacts of Food Choices in the United States." *Environmental Science and Technology* 42: 3508–13.

Weinstein, N., A. K. Przybyiski, and R. M. Ryan. 2009. "Can Nature Make Us More Caring? Effects of Immersion in Nature on Intrinsic Aspirations and Generosity." *Personaility and Social Psychology Bulletin* 35:1315–29.

INDEX

Page numbers in italics refer to maps; numbers in bold refer to figures that appear in the gallery.